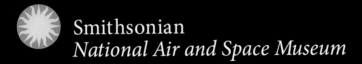

Smithsonian
National Air and Space Museum

MILESTONES OF
FLIGHT

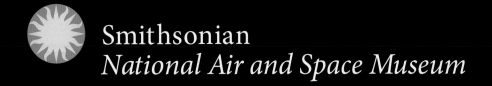

Smithsonian
National Air and Space Museum

MILESTONES OF
FLIGHT

THE EPIC OF AVIATION WITH
THE NATIONAL AIR AND SPACE MUSEUM

F. ROBERT VAN DER LINDEN | ALEX M SPENCER | THOMAS J. PAONE

ZENITH
PRESS

CONTENTS

1. 1903 Wright Flyer 8

2. 1909 Wright Military Flyer and 1911 Wright EX *Vin Fiz* . . . 14

3. Blériot XI and Curtiss D-III Headless Pusher 22

4. Caudron G.4 30

5. Douglas World Cruiser DWC-2 *Chicago* 36

6. Ryan NY-P *Spirit of St. Louis* 44

7. Lockheed Vega 5B and Lockheed Vega 5C *Winnie Mae* 52

8. Piper J-3 Cub 62

9. *Explorer II* 70

10. Douglas DC-3. 76

11. Douglas SBD-6 Dauntless. 82

12. North American P-51D Mustang 90

13. Bell XP-59A Airacomet. 100

14. Messerschmitt Me 262A 1-a Schwalbe (Swallow) 108

15. Boeing B-29 Superfortress *Enola Gay* 118

16. *Little Gee Bee* 128

17. Bell XS-1/X-1 134

18. North American F-86A Sabre 142

19. Boeing 367-80 148

20. McDonnell F-4S-44 Phantom II 154

21. North American X-15 160

22. Arlington Sisu 1A 168

23. Bell UH-1H Iroquois "Huey" *Smokey III* 174

24. Lockheed SR-71A Blackbird 180

25. Aérospatiale-BAC Concorde 190

26. General Atomics MQ-1L Predator 198

Image Credits 204

Index 205

1903 WRIGHT FLYER

CHAPTER 1

BY PETER L. JAKAB

Orville and Wilbur Wright inaugurated the aerial age on December 17, 1903, with their successful first flights of a heavier-than-air flying machine at Kitty Hawk, North Carolina. Their airplane, known as the Wright Flyer, sometimes referred to as the Kitty Hawk Flyer, was the product of a sophisticated four-year program of research and development conducted by the Wright brothers beginning in 1899. Their seminal accomplishment not only encompassed the breakthrough first flight of an airplane, but during the design and construction of their experimental aircraft, the Wrights also pioneered many of the basic tenets and techniques of modern aeronautical engineering, such as the use of a wind tunnel and flight testing as design tools.

The Wright brothers gained an interest in flight as youngsters. In 1878, their father gave them a toy flying helicopter powered by strands of twisted rubber. They played and experimented with it extensively and even built several larger copies of the device. They also had some experience with kites. But it was not until 1896, prompted by the widely publicized fatal crash of famed glider pioneer Otto Lilienthal, that the Wrights began serious study of flight. After absorbing the locally available materials related to the subject, Wilbur wrote to the Smithsonian Institution on May 30, 1899, requesting any publications on aeronautics that it could offer.

Shortly after their receipt of the Smithsonian materials, the Wrights built their first aeronautical craft, a 5-foot-wingspan biplane kite, in the summer of 1899. They chose to follow Lilienthal's lead of using gliders as a steppingstone toward a practical powered airplane, and the 1899 kite was built as a preliminary test device to establish the viability of the control system that they planned to use in their first full-size glider. This means of control would be a central feature of the later successful powered airplane.

The "cockpit" of the Wright Flyer. The hip cradle provided lateral control while the vertical lever just to the left of the hip cradle controlled the elevator for climb and descent.

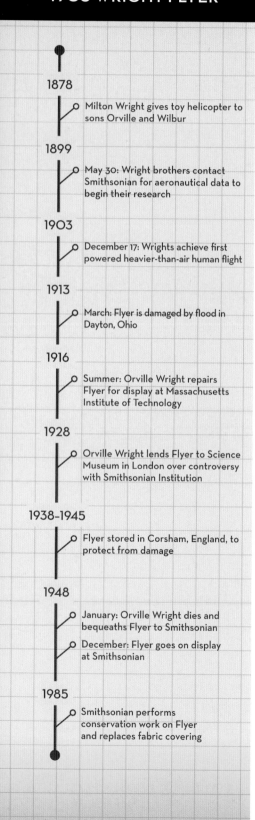

TIMELINE
1903 WRIGHT FLYER

1878

Milton Wright gives toy helicopter to sons Orville and Wilbur

1899

May 30: Wright brothers contact Smithsonian for aeronautical data to begin their research

1903

December 17: Wrights achieve first powered heavier-than-air human flight

1913

March: Flyer is damaged by flood in Dayton, Ohio

1916

Summer: Orville Wright repairs Flyer for display at Massachusetts Institute of Technology

1928

Orville Wright lends Flyer to Science Museum in London over controversy with Smithsonian Institution

1938–1945

Flyer stored in Corsham, England, to protect from damage

1948

January: Orville Wright dies and bequeaths Flyer to Smithsonian

December: Flyer goes on display at Smithsonian

1985

Smithsonian performs conservation work on Flyer and replaces fabric covering

Rather than controlling the craft through altering the center of gravity by shifting the pilot's body weight, as Lilienthal had done, the Wrights intended to balance their glider aerodynamically. They reasoned that if a wing generates lift when presented to an oncoming flow of air, producing differing amounts of lift on either end of the wing would cause one side to rise more than the other, which in turn would bank the entire aircraft. A mechanical means of inducing this differential lift would provide the pilot with effective lateral control. The Wrights accomplished this by twisting, or warping, the tips of the wings in opposite directions via a series of lines attached to the outer edges of the wings and manipulated by the pilot.

Control in climb and descent—or *pitch*, to use the modern term—was also achieved aerodynamically. Similar to his method for lateral balance, Lilienthal had swung his legs fore and aft to balance his glider in pitch, altering the center of gravity with the weight of his legs. The Wrights achieved the same result with a moveable horizontal surface that controlled the movement of the wing's center of lifting pressure fore and aft of a fixed center of gravity. The Wrights' ideas for control advanced aeronautical experimentation significantly because they provided an effective method of controlling an airplane in three-dimensional space and, because they were aerodynamically based, did not limit the size of the aircraft as shifting body weight obviously did. The satisfactory performance of the brothers' 1899 kite demonstrated the practicality of their wing-warping lateral balance and pitch control methods.

Encouraged by this success, the brothers built and flew two full-size piloted gliders in 1900 and 1901. Beyond the issue of control, the Wrights had to grapple with developing an efficient airfoil shape and solving fundamental problems of structural design. Like the kite, these gliders were biplanes. For control of climb and descent, they had forward-mounted horizontal stabilizers. Neither craft had a tail.

The Wrights' home of Dayton, Ohio, did not offer suitable conditions for flying the gliders. An inquiry with the US Weather Bureau identified Kitty Hawk, with its sandy, wide-open spaces and strong, steady winds, as an optimal test site. In September 1900, the Wrights made their first trip to the little fishing hamlet that they would make famous.

With the 1900 glider, although the control system worked well and the craft's structural design was sound, the lift was substantially less than the Wrights' earlier calculations had predicted. In 1901, they returned to Kitty Hawk with a similar glider with increased wing area, hoping this would address the lift deficiency. They also changed the curvature of the wing profile. Not only was the lift even less with the 1901 glider, but problems with the control system they had seemingly proven in 1900 now emerged. Even though the Wrights had achieved greater flight performance with these first two gliders than previous experimenters, they were not satisfied. When they returned from Kitty Hawk in late August 1901, the Wrights were at a low point in their path to a successful airplane.

The brothers began to seriously question the aerodynamic data they had acquired from others, Lilienthal in particular. Now at a critical juncture, Wilbur and Orville decided to conduct an extensive series of tests of wing shapes, building a small wind tunnel in the fall of 1901 to gather a body of accurate aerodynamic data with which to design their next glider. The heart of the Wright wind tunnel was the ingeniously designed pair of test instruments mounted inside to measure coefficients of lift and drag on small model wing shapes, the terms in the equations for calculating lift and drag about which the brothers were in doubt. Others had used wind tunnels before, but no tunnel before the Wrights' was used to gather specific data directly incorporated in the design of an aircraft. From this perspective, the Wrights' wind tunnel research was a breakthrough in aeronautical engineering. In

rudimentary form, the way the Wrights used a wind tunnel remains at the foundation of modern aerodynamic research and design.

The Wrights' third glider, built in 1902 and based on the wind tunnel experiments, was a dramatic success. The lift problems were solved, and with a few refinements to the control system (the key one being a movable vertical tail), they were able to make numerous extended controlled glides. They made between seven hundred and one thousand flights in 1902. The single best one was 622.5 feet in twenty-six seconds. The brothers were now convinced that they stood at the threshold of realizing mechanical flight.

During the spring and summer of 1903, the Wrights built their first powered airplane. Essentially a larger and sturdier version of the 1902 glider, its only fundamentally new component was the propulsion system. With the assistance of their bicycle-shop mechanic, Charles Taylor, the Wrights built a small, 12-horsepower gasoline engine. While the engine was a significant enough achievement, the genuinely innovative feature of the propulsion system was the propellers, which the brothers conceived as rotary wings producing a horizontal thrust force aerodynamically. By turning an airfoil section on its side and spinning it to create an air flow over the surface, the Wrights reasoned that it would generate a horizontal lift force that would propel the airplane forward—thrust. The concept was one of the most original and creative aspects of the Wrights' aeronautical work. The 1903 airplane was fitted with two propellers mounted behind the wings and connected to the engine, which was centrally located on the bottom wing, via a chain-and-sprocket transmission system.

By the fall of 1903, the powered airplane was ready for trial. The Wrights arrived at Kitty Hawk in September with the craft they called simply the Flyer. However, a number of serious problems with the engine transmission system delayed the first flight attempt until mid-December. After winning the toss of a coin to determine who would make the first try, Wilbur took the pilot's position and made an unsuccessful attempt on December 14, damaging the Flyer slightly. Repairs were completed for a second attempt on December 17; it was now Orville's turn. At 10:35 a.m. the Flyer lifted off the beach at Kitty Hawk for a twelve-second flight, traveling 120 feet. Three more flights were made that morning, the brothers alternating as pilot. The second and third ranged between 175 and 200 feet and were fifteen seconds in

Above: Wilbur (left) and Orville Wright pose for the camera on the back porch of their Dayton, Ohio, home, at the height of their fame in June 1909.

Top: The "moment" of invention. The Wright Flyer lifts off at 10:35 a.m. on December 17, 1903, at Kitty Hawk, with Orville Wright at the controls and Wilbur Wright watching alongside.

Top left: The original 1903 Wright Flyer on display in the Smithsonian National Air and Space Museum. The Wright Flyer embodied all the essential design elements of every successful airplane that followed.

Top right: Every aspect of the Wright Flyer was carefully engineered—the aerodynamics, controls, structure, and propulsion system. The Wright brothers not only designed and built the first true airplane, they invented the practice of aeronautical engineering.

duration. With Wilbur at the controls, the fourth and last flight covered 852 feet in fifty-nine seconds. With this final sustained effort, there was no question the Wrights had flown.

As the brothers and the witnesses present discussed the long flight, a gust of wind overturned the Wright Flyer and sent it tumbling across the sand. The aircraft was severely damaged and never flew again; the Wrights crated it and shipped it back to Dayton where it remained in storage in a shed behind their bicycle shop, untouched for more than a decade. But they had successfully demonstrated their design for a heavier-than-air flying machine. They built refined versions of the Flyer in 1904 and 1905, tested at a cow pasture near their home in Dayton, bringing the design to practicality. On October 5, 1905, with the brothers' third powered airplane, Wilbur made a spectacular thirty-nine-minute flight that covered 24.5 miles over a closed course.

In March 1913, nearly a year after Wilbur's death, Dayton was hit by a serious flood, during which the boxes containing the stored Flyer were submerged in water and mud for eleven days. The airplane was uncrated, for the first time since Kitty Hawk, in the summer of 1916, when Orville repaired and reassembled it for a brief exhibition at the Massachusetts Institute of Technology. Several other displays followed at the New York Aero Show in 1917, a Society of Automotive Engineers meeting in Dayton in 1918, the New York Aero Show in 1919, and the National Air Races in Dayton in 1924. On each occasion the Wright Flyer was prepared and assembled for exhibition by a Wright Company mechanic named Jim Jacobs, working under the supervision of Orville.

In 1928 the airplane was placed on loan to the Science Museum in London in response to a controversy between Orville and the Smithsonian Institution concerning misleading and inaccurate claims the Smithsonian made with regard to the aeronautical accomplishments of Smithsonian Secretary Samuel P. Langley. Before shipping it to Europe, Orville and Jim Jacobs refurbished the Flyer extensively; the fabric covering was replaced completely with new material, although it was of the same type as the original Pride of the West muslin. The remaining 1903 fabric that was on the airplane when it flew was saved, and portions of it still exist in various places, including in the National Air and Space Museum collection. During

World War II, the airplane was kept in an underground storage facility near the village of Corsham, approximately 100 miles from London, where various British national treasures were secured. The Flyer was not stored in the London subway as has been often asserted.

In 1942, in a statement published in its annual report, the Smithsonian publicly clarified its position on Langley and unequivocally credited the Wright brothers with the invention of the airplane. Satisfied, Orville privately contacted the Science Museum in 1943 requesting the return of the Flyer to the United States. However, the aircraft remained in safe storage in England until the conclusion of World War II to ensure safe transport. After the war, the Science Museum requested an extension of the loan from Orville in order to build an accurate reproduction of the Flyer. Orville agreed. Upon his death of a heart attack in January 1948, the original 1903 Wright Flyer was bequeathed to the Smithsonian Institution. The airplane was returned to the United States in November 1948 and formally installed at the Smithsonian in an elaborate ceremony on December 17, the forty-fifth anniversary of the first flights. It has been on public display at the Smithsonian ever since.

The Flyer received some minor repairs and cleaning in 1976 just before being moved into the Smithsonian's then-new National Air and Space Museum building. In 1985, it was given its first major treatment since preparing it for loan to the Science Museum in late 1926 and early 1927. It was completely disassembled, the parts thoroughly cleaned and preserved, and all new fabric covering applied. A careful search was made to locate new fabric that matched the original as closely as possible. When the fabric was replaced in 1927, it was sewn on in a slightly different way than originally done by the brothers in 1903. When stitching the new fabric in 1985, a large section of original-flown 1903 wing covering in the National Air and Space Museum collection was used as a pattern, ensuring the accuracy of the 1985 restoration.

The original 1903 Wright Flyer is truly a seminal artifact. All subsequent airplanes embody the fundamental elements of its design, and it is not too much to say this object changed the world. Moreover, the Wrights invented aeronautical engineering in the modern sense, an equally revolutionary achievement.

Bottom: The official installation of the Wright Flyer at the Smithsonian Institution. An elaborate ceremony was held on December 17, 1948, the forty-fifth anniversary of the historic flights at Kitty Hawk.

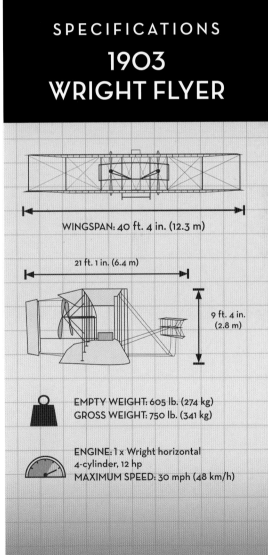

SPECIFICATIONS
1903 WRIGHT FLYER

WINGSPAN: 40 ft. 4 in. (12.3 m)

21 ft. 1 in. (6.4 m)

9 ft. 4 in. (2.8 m)

EMPTY WEIGHT: 605 lb. (274 kg)
GROSS WEIGHT: 750 lb. (341 kg)

ENGINE: 1 x Wright horizontal 4-cylinder, 12 hp
MAXIMUM SPEED: 30 mph (48 km/h)

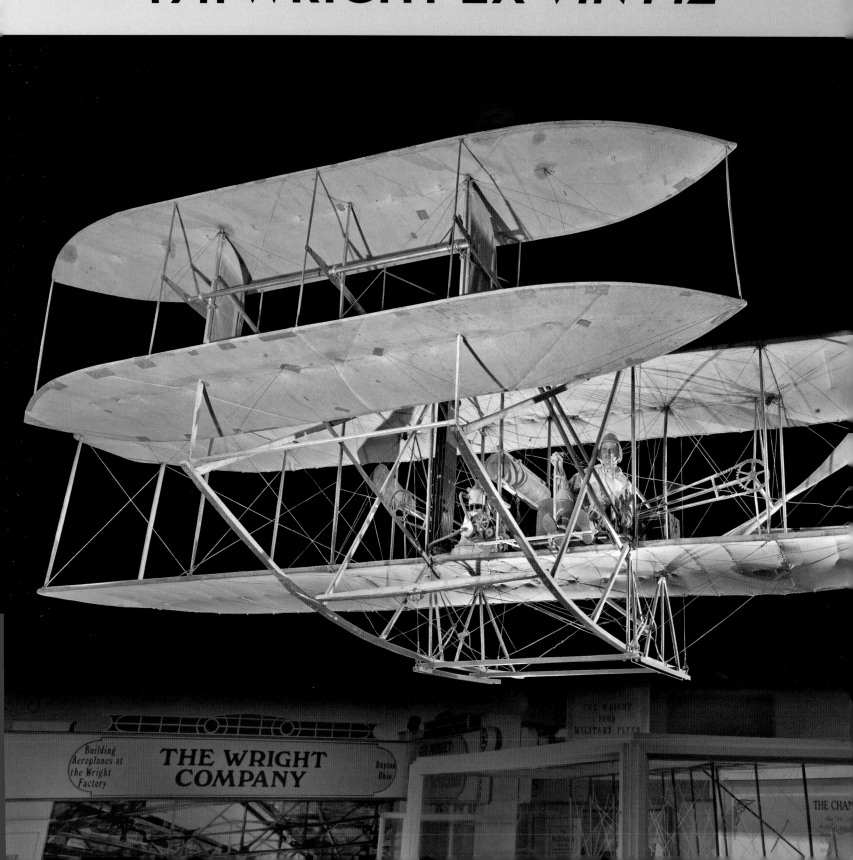

Building
Aeroplanes at
the Wright
Factory

THE WRIGHT
COMPANY

Dayton
Ohio

CHAPTER 2

BY PETER L. JAKAB

In 1908, the US Army Signal Corps advertised for bids for a two-seat observation aircraft. The general requirements were that it be designed for easy assembly and disassembly so that an army wagon could transport it; that it be able to carry two people with a combined weight of 350 pounds and sufficient fuel for 125 miles; and that it be able to reach a speed of at least 40 miles per hour in still air, to be calculated during a two-lap test flight over a 5-mile course, with and against the wind. In addition, the Signal Corps stipulated that the design must demonstrate the ability to remain in the air for at least one hour without landing, and then land without causing any damage that would prevent it from immediately starting another flight. It would also need be able to ascend in any sort of terrain in which the Signal Corps might need it in field service and land without requiring a specially prepared spot; be able to land safely in case of accident to the propelling machinery; and be simple enough to permit someone to become proficient in its operation within a reasonable amount of time.

The purchase price was set at $25,000 with 10 percent added for each full mile per hour of speed over the required 40 miles per hour and 10 percent deducted for each full mile per hour under 40 miles per hour.

In response to the army's advertisement for bids, the Wright brothers constructed a two-seat, wire-braced biplane

The 1909 Wright Military Flyer on display in the Smithsonian National Air and Space Museum. The world's first military airplane has been in the Smithsonian collection since 1911.

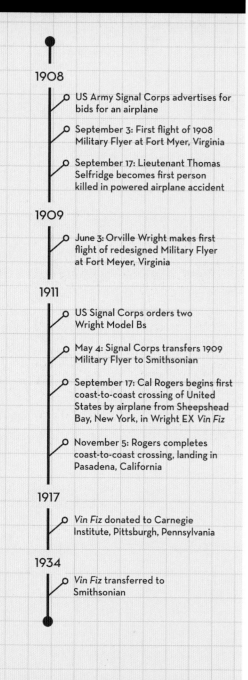

TIMELINE

1909 WRIGHT MILITARY FLYER AND 1911 EX *VIN FIZ*

1908

- US Army Signal Corps advertises for bids for an airplane
- September 3: First flight of 1908 Military Flyer at Fort Myer, Virginia
- September 17: Lieutenant Thomas Selfridge becomes first person killed in powered airplane accident

1909

- June 3: Orville Wright makes first flight of redesigned Military Flyer at Fort Meyer, Virginia

1911

- US Signal Corps orders two Wright Model Bs
- May 4: Signal Corps transfers 1909 Military Flyer to Smithsonian
- September 17: Cal Rogers begins first coast-to-coast crossing of United States by airplane from Sheepshead Bay, New York, in Wright EX *Vin Fiz*
- November 5: Rogers completes coast-to-coast crossing, landing in Pasadena, California

1917

- *Vin Fiz* donated to Carnegie Institute, Pittsburgh, Pennsylvania

1934

- *Vin Fiz* transferred to Smithsonian

with a 30- to 40-horsepower Wright vertical 4-cylinder engine driving two wooden propellers. With Wilbur in Europe, where he had been demonstrating a similar airplane in France in the summer of 1908, Orville made the first flight of this airplane at Fort Myer, Virginia, on September 3. Several days of very successful and increasingly ambitious flights followed. Orville set new duration records day after day, including a seventy-minute flight on September 11. He also made two flights with a passenger.

On September 17, however, tragedy struck. At 5:14 p.m., Orville took off with Lt. Thomas E. Selfridge, the army's observer, as his passenger. The airplane had circled the field four and a half times when a propeller blade split. The aircraft, then at 150 feet, safely glided to 75 feet before plunging to earth. Orville was severely injured, including facial lacerations, a broken hip, a broken leg, and four broken ribs. Lieutenant Selfridge, however, died later that evening from a fractured skull. The aircraft was destroyed, and Selfridge became the first fatality in a powered airplane accident.

On June 3, 1909, the Wrights returned to Fort Myer with a new machine to complete the trials begun in 1908. The brothers agreed that even though Wilbur was present this time, Orville would do all the flying and complete what he had started the previous year. The engine was the same as in the earlier aircraft, but the 1909 model had a smaller wing area and modifications to the rudder and the wire bracing. Lieutenant Frank P. Lahm and Lt. Benjamin D. Foulois, future army pilots, were the Wrights' passengers.

On July 27, with Lahm on board, Orville made a record flight of one hour, twelve minutes, forty seconds, covering approximately 40 miles, satisfying the army's endurance and passenger-carrying requirements. To establish the speed of the airplane, a course was set up from Fort Myer to Shooter's Hill in Alexandria, Virginia, a distance of 5 miles. After waiting several days for optimum wind conditions, Orville and Foulois made the 10-mile round trip on July 30. The out-lap speed was 37.7 miles per hour and the return lap was 47.4 miles per hour, for an average speed of 42.5 miles per hour. For the 2 miles per hour over the required 40, the Wrights earned an additional $5,000, making the final sale price of the airplane $30,000.

Upon taking possession of the Military Flyer (referred to as the Signal Corps No. 1 by the War Department), the army conducted flight training at nearby College Park, Maryland, beginning in October 1909 and at Fort Sam Houston in San Antonio, Texas, in 1910. Various modifications were made to the Military Flyer during this period, the most significant being the addition of wheels to the landing skids.

The brothers had formed the Wright Company in late 1910 to begin manufacture and sale of airplanes. Their first commercial design, the Model B, was the first Wright airplane with wheeled landing gear and the horizontal stabilizer/elevator at the rear. Early in 1911, the Signal Corps placed an order for two of the new Model B airplanes. In addition, the War Department proposed shipment of the original 1909 Signal Corps airplane to the Wright Company factory in Dayton, Ohio, for rebuilding with Model B controls and other improvements. The Wright Company quoted a price of $2,000 for the upgrade but advised against it because of the many design improvements that had been made during the intervening two years.

The manager of the Wright Company, Frank Russell, learned that the Smithsonian Institution was interested in the first army airplane and would welcome its donation. The War Department agreed and approved the transfer on May 4, 1911. The aircraft was

The 1909 Wright Military Flyer during the flight trials at Fort Myer, Virginia, on July 27, 1909. Orville Wright is in the foreground; Wilbur Wright is wearing the straw hat.

restored close to its original 1909 configuration, but a few non-original braces added for the wheeled landing gear in 1910 remained on the airplane when it was turned over to the Smithsonian. Apart from a few minor repairs over the years, the airplane has not been restored since its acquisition in 1911. Of the three Wright airplanes in the National Air and Space Museum collection, the 1909 Military Flyer retains the largest percentage of its original material and components.

In 1911, Calbraith Perry Rodgers achieved the first crossing of the United States by airplane in his Wright EX biplane, named the *Vin Fiz*. Rodgers had a rich personal heritage of exploration and adventure: he was a descendant of Cdre. Mathew Calbraith Perry, famous for opening Japan to US trade in 1853; Perry's brother, Oliver Hazard Perry, who led an important naval victory during the War of 1812; and Cdre. John Rodgers, who also figured prominently in the War of 1812. However, Rodgers's interest in attending the US

Naval Academy was thwarted by his deafness, a condition resulting from a serious bout with scarlet fever as a young boy. Given his nautical lineage, he was an avid sailor and was elected to the prestigious New York Yacht Club in 1902. He pursued his love of speed through the recently introduced technologies of motorcycles and automobiles.

Rodgers, typically called Cal, was introduced to aviation in June 1911. His cousin, John, a Naval Academy graduate, had been selected to learn to fly the navy's newly purchased Wright airplane and was sent to the Wright factory and flying school in Dayton, Ohio, in March 1911. On a visit to his cousin during his training, Cal was immediately hooked on flying and soon began flight instruction. Shortly thereafter, with his cousin, he ordered a new Wright Model B airplane. A quick study, Rodgers was flying public exhibitions in Ohio and Indiana with his new aircraft by July. On August 7, 1911, he passed his flight tests for pilot's license number 49, issued by the Aero Club of America, and on August 10 he arrived in Chicago to compete in the Chicago International Aviation Meet at Grant Park. He won the duration contest, and along with his performance in other events he earned total prize money of $11,285 and instant celebrity.

Upon completing his transcontinental flight from New York to California, Cal Rodgers rolled the 1911 Wright EX *Vin Fiz* into surf of the Pacific Ocean for a photo op.

Ten months earlier, famed publishing magnate William Randolph Hearst had captured the attention of the aviation world when he announced a $50,000 prize for the first flight across the United States in thirty days or less. The offer was good for one year beginning on October 10, 1910. The bold challenge interested many of the leading names in aviation, including the Wright brothers and Glenn Curtiss, but the technical and logistical demands of such a flight precluded any immediate attempts.

In September 1911, three competitors were finally in the race: Robert Fowler, James Ward, and Cal Rodgers. Fowler took off from San Francisco on September 11 but aborted his attempt by the end of the month after three failed attempts to cross the Sierras. Ward took off from the East Coast on September 13 but withdrew little more than a week later, not even making it out of New York State.

While still in Chicago, Rodgers secured financial backing for his coast-to-coast attempt from the Armour Company, a local firm then introducing a new grape-flavored soft drink called Vin Fiz. Armour provided a special train, emblazoned with the Vin Fiz logo and including cars to accommodate Rodgers's family and support crew, as well as a "hangar" car filled with spare parts to repair and maintain the airplane over the course of the flight. There was even an automobile on board to pick up Rodgers after forced landings away from the rail line. The pilot would receive $5 for every mile he flew east of the Mississippi River and $4 for every mile west of the river. Rodgers agreed to pay for the fuel, oil, spare parts, mechanics, and the airplane itself, which the Wright Company agreed to build. Chief mechanic on the flight was Charles Taylor, who had worked for the Wright brothers since 1901 and built the engine for the 1903 Wright Flyer.

Rodgers's airplane was a Wright EX, a special design used for exhibition flying that was a slightly smaller, single-seat version of the Wright Company's standard Model B. Like the support railcar, it carried the Vin Fiz logo on its wings and tail and was quickly dubbed the Wright EX *Vin Fiz*. It was powered by a 35-horsepower Wright vertical 4-cylinder engine, and it carried enough fuel for a maximum of three and a half hours of flying.

Rodgers began his epic journey from Sheepshead Bay, New York, on September 17, 1911. The flight was punctuated by numerous stops, delays, and accidents. When the thirty-day time limit Hearst imposed for the $50,000 prize had expired, he had only reached Kansas City, Missouri. Undaunted, Rodgers continued on, determined to make the first transcontinental airplane flight whether he received the money or not. Upon leaving Kansas City, he flew due south to Texas and then made his way along the southern US border toward Pasadena, California, the flight's official termination point.

Rodgers continued to experience frequent mechanical failures, damage to the airplane in hard landings, and weather delays. More serious trouble arose on November 3, shortly after passing over Imperial Junction, California, less than 200 miles from the finish. At 4,000 feet, an engine cylinder exploded, damaging one of the wings and driving steel shards into Rodgers's right arm. He struggled to regain control of the *Vin Fiz* and, amazingly, managed to glide the 6 miles back to Imperial Junction and land safely. The engine and airplane were repaired a day later, and despite his painful injury, Rodgers departed for Pasadena once again on November 4. Further engine problems forced him down at Banning, California, about halfway to his final destination. On November 5, he was airborne again, and after brief stops in Beaumont and Pomona, he arrived in Pasadena to a hero's welcome—forty-nine days after setting out from Sheepshead Bay.

SPECIFICATIONS
1909 WRIGHT MILITARY FLYER

WINGSPAN: 36 ft. 8 in. (11.2 m)

29 ft. 2 in. (8.9 m)

8 ft. 2 in. (2.5 m)

EMPTY WEIGHT: 735 lb. (334 kg)

ENGINE: 1 x Wright vertical 4-cylinder, 30–40 hp
MAXIMUM SPEED: 42 mph (68 km/h)

The 1911 Wright EX *Vin Fiz* on display in the Smithsonian National Air and Space Museum. The Armor Company, producer of the grape drink Vin Fiz, sponsored the flight.

Although Pasadena was the official end of the coast-to-coast journey, Rodgers wanted to fly all the way to the Pacific shore. Several coastal towns bid for the honor, and Long Beach was the final selection. They agreed to pay Rodgers $1,000, plus part of the gate receipts of an exhibition of the *Vin Fiz* after arrival. Rodgers took off from Pasadena for the short 23-mile trip on November 12. Minutes into the flight, however, engine failure forced him down near Covina Junction. Repairs to a broken fuel line had him back in the air that afternoon. But yet again, just a short time after takeoff, near Compton, Rodgers was down. This time it was a serious crash; the airplane was severely damaged and Rodgers badly hurt. It took several weeks to make the *Vin Fiz* airworthy again and for Rodgers to recover from his injuries.

On December 10, needing crutches to move about on his still-healing ankle, Rodgers boarded his battered aircraft, determined to fly the *Vin Fiz* all the way to the Pacific Ocean. As at Pasadena, his arrival was an organized public event and a large crowd gathered at Long Beach to welcome him. The remainder of the great adventure met without incident, and Rodgers landed to cheering crowds. To create a photo opportunity for the press and spectators, the *Vin Fiz* was rolled into the surf, allowing the Pacific to lap over it wheels. The 4,321 miles had been covered in eighty-two hours and four minutes total flying time at an average speed of 51.5 miles per hour. Cal Rodgers had secured his place in aviation history. (Robert Fowler began another west-to-east transcontinental flight on October 19, this time taking a southern route to avoid the mountains. He arrived in Jacksonville, Florida, on February 8, 1912, completing the trip in more than twice as many days as Rodgers.)

Rodgers and his wife liked Pasadena and decided to stay. He kept the *Vin Fiz* and a second airplane, a two-seat Wright Model B, at nearby Dominguez Field, making exhibition flights with the *Vin Fiz* and taking up passengers and giving flight instruction with the Model B. He later moved the airplanes to Long Beach and operated from there. On April 3, 1912, Rodgers was airborne in the Model B, making a test flight after some engine tuning. Witnesses observed a steep dive as Rodgers apparently attempted to avoid a flock of seagulls. In the next instant he was seen struggling with the controls just before the airplane crashed into the surf, only 100 yards from his landing spot for the last leg of the transcontinental flight in December. He was killed instantly. Various causes for the accident were put forth, ranging from a seagull jamming the controls to Rodgers's recklessness or carelessness as a pilot. The precise cause remains undetermined. The wreckage of Rodgers's Model B was acquired by one of his mechanics, Frank Shaffer, and his partner Jesse Brabazon. They rebuilt it and flew it for approximately one year until it was destroyed in a crash while piloted by Brabazon's friend, Andrew Drew, who was killed.

Cal's cousin, Lt. John Rodgers, acquired the *Vin Fiz* after Cal's death and offered it to the Smithsonian Institution, but it was not accepted on the grounds that it was very similar to the recently acquired 1909 Wright Military Flyer. It then passed to Rodgers's wife, Mabel, who, not long after Cal's death, married Charles Wiggin. They exhibited and flew the *Vin Fiz* publicly for two years until Rodgers's mother was awarded possession of the airplane in 1914 in a court ruling regarding his estate. According to Charles Taylor, the Wright mechanic who assisted Rodgers on the transcontinental flight, Rodgers's mother shipped the *Vin Fiz* to the Wright factory in Dayton, Ohio, for refurbishment but was either unable or unwilling to pay for the work, allowing the airplane to languish at the Wright factory until it was destroyed in 1916 after the company was sold. In reality, Rodgers's mother had the *Vin Fiz* restored and donated it to the Carnegie Institute in Pittsburgh in 1917. The airplane was later acquired by the Smithsonian from Carnegie in 1934.

The probable explanation for the conflicting information lies in the misconception that there was a single *Vin Fiz* airframe. On the transcontinental flight, several sets of wings and a large supply of other components and spare parts were brought along on the support train. Rodgers's airplane was repaired and rebuilt many times during the trip. By the time the wheels of the *Vin Fiz* were rolled into the Pacific at Long Beach, almost nothing of the airplane that took off from Sheepshead Bay remained. As a result, at the end of the journey, there were enough flown, genuine *Vin Fiz* parts to make more than one airplane. Charles Taylor was probably accurate when he stated that the *Vin Fiz* sent to the Wright factory (i.e., the intact airplane that Mabel and Charles Wiggin flew in 1912–1914) was destroyed. The airplane that ended up at the Carnegie Institute, and then the Smithsonian, was very likely reconstructed from the parts left over from the many repairs and rebuilds during the flight. Thus, the airplane in the National Air and Space Museum collection is genuine in that it comprises components that, at various points, were part of the *Vin Fiz* during its historic coast-to-coast flight.

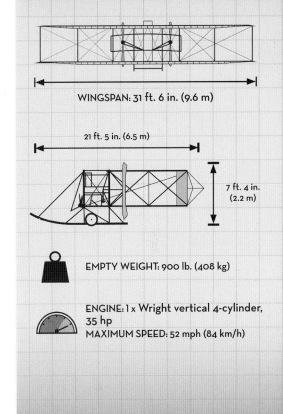

WINGSPAN: 31 ft. 6 in. (9.6 m)

21 ft. 5 in. (6.5 m)

7 ft. 4 in. (2.2 m)

EMPTY WEIGHT: 900 lb. (408 kg)

ENGINE: 1 x Wright vertical 4-cylinder, 35 hp
MAXIMUM SPEED: 52 mph (84 km/h)

BLÉRIOT XI AND CURTISS D-III HEADLESS PUSHER

CHAPTER 3

BY TOM D. CROUCH

The sun was barely up on the morning of July 25, 1909, when thirty-seven-year-old Louis Blériot landed his frail wood-and-fabric monoplane on English soil. He had covered the roughly 23 miles between his takeoff point near Calais, France, and his destination of Dover, England, in just over thirty-six minutes, and the short, early-morning flight earned the daring pilot a £1,000 prize offered by the London *Daily Mail* for the first flight across the English Channel.

The deeper meaning of the achievement was apparent. As the *Pall Mall Gazette* observed, Blériot had created "a revolution in human affairs." In less than half an hour he had "brought home to the minds of the people of all nations the possibilities of the flying machine in the future." Lord Northcliffe, the English press baron who had offered the prize, summed the matter up more succinctly, noting, "England is no longer an island."

The airplane that made the historic flight was the most basic flying machine imaginable. With its open box-frame fuselage, exposed pilot position, and 3-cylinder engine producing all of 25 horsepower, it was scarcely more of a threat than the balloons that preceded it. Yet for all of its apparent fragility, the classic Blériot Type XI monoplane was the most significant

The John Domenjoz Blériot XI on display at the Smithsonian National Air and Space Museum. The Museum's Paul E. Garber purchased it in 1950—along with two other aircraft—all for $2,500.

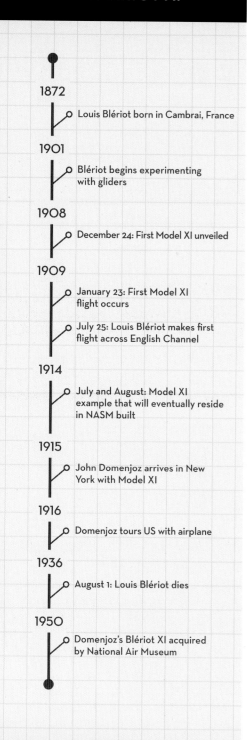

1872
- Louis Blériot born in Cambrai, France

1901
- Blériot begins experimenting with gliders

1908
- December 24: First Model XI unveiled

1909
- January 23: First Model XI flight occurs
- July 25: Louis Blériot makes first flight across English Channel

1914
- July and August: Model XI example that will eventually reside in NASM built

1915
- John Domenjoz arrives in New York with Model XI

1916
- Domenjoz tours US with airplane

1936
- August 1: Louis Blériot dies

1950
- Domenjoz's Blériot XI acquired by National Air Museum

and influential aircraft design of the post-Wright era. Between July 1909 and August 1914, as many as eight hundred Blériot aircraft took to the sky, most of them Type XI derivatives. A 1910 company sales brochure listed the names of 104 individuals who had purchased Type XI machines and proudly noted that the embryonic air forces of France, Britain, Italy, Austria, and Russia were flying their aircraft. Flying clubs as distant as Saigon and Sebastopol were operating Blériots as well. In 1911, aviators Claude Grahame-White and Harry Harper listed 158 Type XIs in the hands of private pilots and noted that it was the most popular monoplane ever produced.

Blériot dominated the air racing and exhibition circuit in the years leading up to World War I. Major speed, distance, and altitude competitions offered by air meets across Europe were captured by pilots flying the classic monoplane. The high-water mark came in July 1910, by which time the Type XI held the world's records for altitude, speed, distance, and duration. As a result of the enormous success of the French-built originals, foreign manufacturers paid a premium for the right to produce Blériots under license. Countless other less exact and usually much less airworthy copies emerged from garages and workshops across Europe and America. Strangely, though, when the original Type XI was first unveiled at the Annual Paris Automobile Salon on Christmas Eve 1908, the authoritative publication *Flight* remarked, "It was not unusual to find doubts expressed as to its capacity for flight at all."

A native of the northern industrial town of Cambrai, Louis Blériot was a striking man with a sweeping moustache, clear, deep-set brown eyes, and high cheekbones that led more than one commentator to remark on his resemblance to an ancient Gallic chieftain. Frederick Collin, his mechanic, called attention to his "patron's" prominent nose, suggesting that such a bird-like profile must be evidence of predestination. Soon after graduating from the École Centrale des Arts et Manufactures he founded La Société des Phares Blériot, a firm specializing in the production of acetylene headlamps and accessories for automobiles. Married to Alice Védères, and with the first of their six children beginning to arrive, the young man seemed to be settling into the position of a prosperous small industrialist.

But Blériot was already infected by the flying bug. He had been interested in the possibility of heavier-than-air flight while still a student but had kept his enthusiasm in check "for fear of being taken for a fool." Now, earning an average of sixty thousand francs a year from the sale of his headlamps, he could afford to indulge in his first aeronautical experiments and was naturally drawn into a circle of young French aeronautical enthusiasts.

Between 1901 and 1909, Blériot experimented with a series of models, gliders, and powered machines, but the Type XI was his first real success. Although Raymond Saulnier, who would soon emerge as a leading manufacturer in his own right, was involved in the design of the machine, it was apparent that Blériot had a good deal to do with engineering the new monoplane. Key features included a three-wheel undercarriage with the front wheels mounted on a shock-absorbing bedstead; pylon wing supports; a rectangular "trellis" fuselage, uncovered at the rear; a small rudder and pivoting elevators; and the classic Blériot-invented stick-and-rudder arrangement still used today.

Blériot borrowed the idea of wing warping for lateral control from the Wright brothers. He was astounded at the degree to which Wilbur Wright controlled his aircraft during his initial demonstration flights in France in 1908. "I consider that for us in France, and everywhere, a

Louis Blériot is seen in the cockpit of the first Type XI.

new era in mechanical flight has commenced," he commented. Ross Browne, an American who had witnessed the scene, remarked that "Blériot was all excited; he looked over the machine . . . he tested the wings, and Mr. Wright showed how the warping was done . . . how it worked." Blériot could scarcely be contained. "I'm going to use a warped wing," he announced to his entourage. "To hell with the aileron." He was, Browne recalled, "just like a young boy."

Blériot flew the Type XI for the first time on January 23, 1909. On July 14, he piloted his machine from Étampes to Orléans, capturing a government subsidized prize of fourteen thousand francs offered by the Aéro-Club de France for the first straight line flight of over 40 kilometers.

The English Channel offered his next challenge. The Northcliffe Prize attracted other competitors, notably Hubert Latham, a native Parisian who had led safaris to Africa, raced automobiles and motorboats, and ballooned across the Channel. Latham had tried and failed the crossing once and was on the ground waiting for better weather when his rival Blériot flew into history.

The Type XI monoplane evolved with the passage of time and the advance of aeronautical technology. The addition of a 50-horsepower Gnome rotary engine as the standard power plant beginning in 1910 increased the aircraft's speed from 40 to 50 miles per hour. A series of deadly crashes led to a redesign of the Blériot's thin wing into a thicker lifting surface with improved bracing. By September 1913, Adolphe Pégoud, a Blériot flying instructor, was looping the loop and giving aerobatic performances in a Type XI specially braced and fitted with a seatbelt and other safety features.

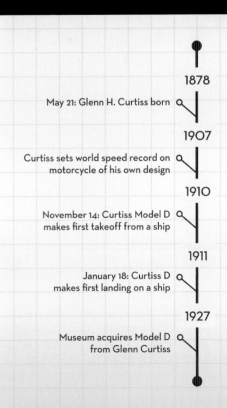

TIMELINE
CURTISS D-III
HEADLESS PUSHER

1878
May 21: Glenn H. Curtiss born

1907
Curtiss sets world speed record on motorcycle of his own design

1910
November 14: Curtiss Model D makes first takeoff from a ship

1911
January 18: Curtiss D makes first landing on a ship

1927
Museum acquires Model D from Glenn Curtiss

The National Air and Space Museum's Type XI was one of the last of the peacetime production airplanes to roll out of the factory in the Paris suburbs in July and August 1914. The aircraft was earmarked for John Domenjoz, a Swiss-born Blériot factory instructor and test pilot who planned to tour Europe with his new machine. With the outbreak of war, however, he decided to pack his airplane and sail for South America, where he thrilled crowds with loops, corkscrew turns, and other daring maneuvers. His exhibitions including inverted flights lasting over a minute, a feat that earned him the sobriquet "Upside Down Domenjoz."

On September 28, 1915, Domenjoz and his plane arrived in New York at the invitation of G. J. Kluyskens, a US dealer for Blériot aircraft and Gnome and Anzani engines. Domenjoz and his aerobatic performances quickly became a major draw at the Sheepshead Bay

This overhead view offers a great look at the Domenjoz Blériot XI after restoration at the Paul E. Garber Facility

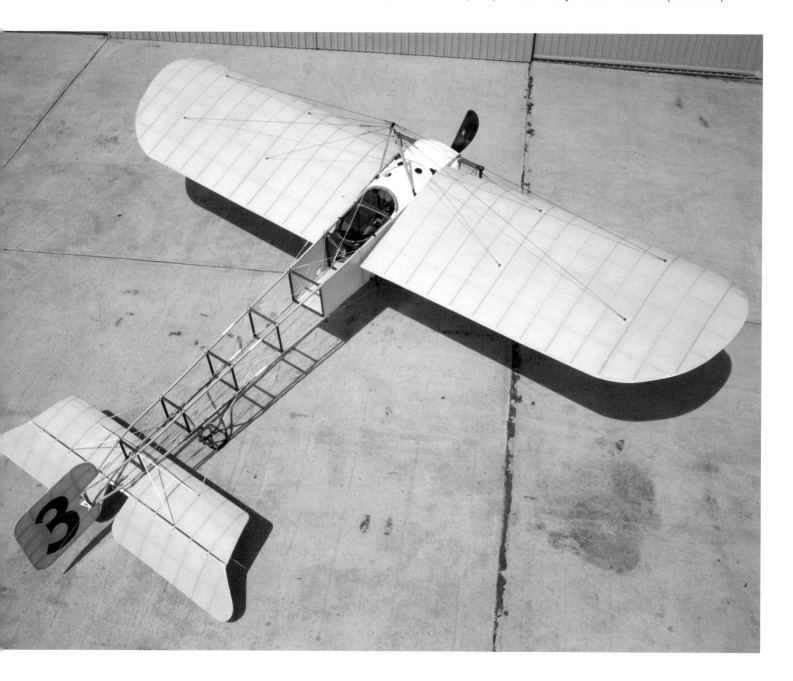

racetrack on Long Island. On October 7, a *New York Times* reporter was on hand to describe the scene as the daring aviator ventured aloft in spite of weather so bad that the auto races had been canceled. "Domenjoz looped the loop, frontward and backward, in his twenty-minute flight," the reporter enthused. "He drove his plane through the air upside down, landing as lightly as a feather on almost the same spot from which he took off."

On October 13, Domenjoz suffered an engine failure while attempting to circle Manhattan but earned the praise of newsmen for safely gliding over a residential district before making a safe landing. A month later, on Election Day, he performed loops, spirals, and "quick and fancy turns" over the Statue of Liberty. Following an appearance in Richmond, Virginia, in February 1916, Domenjoz launched a tour that carried him to Washington, D.C., south to Havana, and back for a return engagement at Sheepshead Bay. That summer he partnered with Curtiss pilot Baxter Adams on a midwestern tour during which the pair demonstrated such military skills as bomb dropping and mock dogfighting.

In the winter of 1916, Domenjoz returned to France, where he worked as a test pilot for the Société Pour L'Aviation et ses Dérivés (SPAD), which a consortium led by Blériot had purchased in 1913. By the spring of 1917 he was back in the United States for another exhibition season, followed by wartime service as a flying instructor at Park Field in Memphis, Tennessee. In 1919, he undertook a final exhibition tour with his now-aging Blériot before putting the airplane in storage in a barn. He returned to France, where he lived for the next seventeen years.

Domenjoz returned again to the United States in 1937 with little hope of ever seeing his airplane again. He was surprised to learn that it had been sold to an aviation museum at Roosevelt Field in Mineola, Long Island. "Domenjoz was practically in tears of joy," a friend recalled, "and left immediately for the field to see his 'beautiful little airplane,' as he called it." There it remained until 1950, when Paul E. Garber of the National Air Museum was able to purchase it, along with a Nieuport scout and Thomas Scott Baldwin's Red Devil biplane—all for $2,500.

Today, John Domenjoz's "beautiful little airplane" is one of the jewels of the National Air and Space Museum's Early Flight collection. Its long and tangled operational history is surely a match for any other surviving late-model Type XI.

If the Blériot XI typifies the best that European aviation had to offer in the years before World War I, surely the Curtiss D-III Headless Pusher qualifies as one of the most popular American-designed and American-built aircraft of that period.

Glenn Hammond Curtiss, a native of Hammondsport, New York, was in love with speed. As a young man he graduated from bicycle and motorcycle racing to the design and construction of motorcycles and engines. In 1907, he set a world speed record of 136.37 miles per hour riding one of his own motorcycles over the sand at Ormond Beach, Florida. The quality of his engines caught the attentions of pioneer airman Thomas Scott Baldwin and, through him, Alexander Graham Bell, who was organizing the Aerial Experiment Association (AEA) to build and fly aircraft. At Bell's suggestion, the group began to employ ailerons for lateral control rather than the wing-warping system favored by the Wrights.

When Curtiss stepped away from the AEA, established the Herring-Curtiss Company, and began to sell airplanes, the Wright brothers brought a patent infringement suit against him that would continue in the courts from 1910 to 1917. Even as the Wrights won every legal judgment, Curtiss trusted his lawyers to keep the case alive and defend him in court.

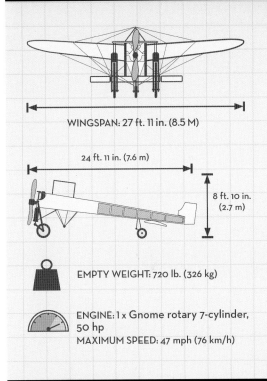

WINGSPAN: 27 ft. 11 in. (8.5 M)

24 ft. 11 in. (7.6 m)

8 ft. 10 in. (2.7 m)

EMPTY WEIGHT: 720 lb. (326 kg)

ENGINE: 1 x Gnome rotary 7-cylinder, 50 hp
MAXIMUM SPEED: 47 mph (76 km/h)

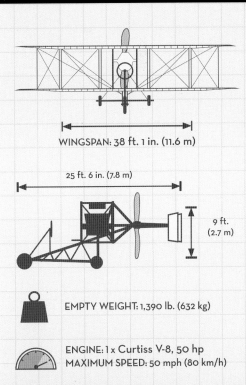

WINGSPAN: 38 ft. 1 in. (11.6 m)

25 ft. 6 in. (7.8 m)

9 ft. (2.7 m)

EMPTY WEIGHT: 1,390 lb. (632 kg)

ENGINE: 1 x Curtiss V-8, 50 hp
MAXIMUM SPEED: 50 mph (80 km/h)

A Curtiss D "headless pusher" biplane. The standard Curtiss design by 1911, it followed the general basic configuration established by the Wrights.

Outside the courts he earned fame by winning a series of national and international prizes and competitions, emerging as the leading American aircraft manufacturer of the prewar era. By 1911, production was focused on the Model D, the standard Curtiss design that followed the general configuration established by the Wrights: a braced biplane with a forward elevator, or canard, and a rear-mounted engine and propeller that pushed the airplane through the air. It was with a Model D that aviator Eugene Ely made the first takeoff from the deck of a ship (November 14, 1910) and the first deck landing (January 18, 1911). It was also the machine in which exhibition pilots hired by Curtiss performed for crowds across the United States.

Over time, Curtiss and his aviators discovered that a single elevator in the rear was preferable to a canard, transforming the aircraft into a Curtiss D Headless Pusher. Like the Blériot XI, the Curtiss D became the design of choice for would-be pilots and amateur builders across America.

Paul Garber, then a Smithsonian assistant curator, saw Curtiss pilot Bert Acosta fly a Curtiss D Headless at the 1925 National Air Races at Mitchell Field, Long Island, and asked Glenn Curtiss if the company would be willing to donate the aircraft to the institution. The machine had apparently been assembled in about 1919 from parts found in the Curtiss factory. After several months of negotiation with company officers, the aircraft arrived at the museum in December 1927. Discovering that his new acquisition had been fitted with a modern control system and engine, Garber returned it to its 1911 configuration with an appropriate engine. It is on view today in the museum's Early Flight Gallery along with the Blériot XI, the Wright Military Flyer, the Ecker flying boat, and other aircraft and objects that invite visitors to share a sense of the wonder and excitement generated by the birth of aviation.

The Smithsonian National Air and Space Museum's Curtiss D Headless Pusher flew at the National Air Races in 1925 and was donated the Museum in 1927. When it was discovered to have been fitted with a modern control system and engine, curator Paul Garber returned it to its 1911 configuration.

CAUDRON G.4

CHAPTER 4

BY PETER L. JAKAB

Gaston and René Caudron were among the earliest aircraft manufacturers in France. After building and testing a few original designs in 1909 and early in 1910, the brothers established a flight training school at Le Crotoy and an aircraft factory at Rue, both in northern France, in 1910.

The first factory-produced Caudron was the Type A4, a 35-horsepower Anzani-powered tractor biplane in which the pilot sat completely exposed behind the rear spar of the lower wing. The next major design, the Type B, was the first to feature the abbreviated fuselage/pilot nacelle characteristic of many later Caudron aircraft. It was powered by a 70-horsepower Gnome or a 60-horsepower Anzani engine mounted in the front of the nacelle with the pilot immediately behind. Although a tractor configuration, the tail unit of the Type B was supported by booms extending from the trailing edge of the wings, an arrangement more commonly featured on pusher aircraft. Lateral control was accomplished with wing warping. The Type B established the basic configuration of Caudron designs through the G.4 model.

The first of the well-known Caudron G series aircraft appeared in 1912. Initially designed as a trainer, the Type G was developed into the G.2 by the outbreak of World War I and saw limited military service in 1914 in single- and

The Caudron G.4 on display at the Smithsonian National Air and Space Museum retains its original 1916 fabric, making it one of only a handful of surviving World War I aircraft still in original condition.

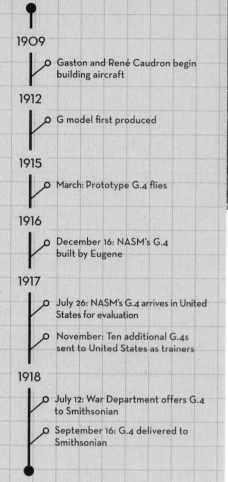

1909

○ Gaston and René Caudron begin building aircraft

1912

○ G model first produced

1915

○ March: Prototype G.4 flies

1916

○ December 16: NASM's G.4 built by Eugene

1917

○ July 26: NASM's G.4 arrives in United States for evaluation

○ November: Ten additional G.4s sent to United States as trainers

1918

○ July 12: War Department offers G.4 to Smithsonian

○ September 16: G.4 delivered to Smithsonian

two-seat versions. By that time the Caudron factory had been relocated to Lyon, where an improved version, designated the G.3, was being produced in significant numbers. Soon a second factory was opened at Issy-les-Moulineaux, near Paris, to meet military demand for the airplane. The G.3 was primarily a two-seat aircraft, but a few were converted to single-seat versions. They were powered variously by 80-horsepower Le Rhône and Gnome rotary engines and 90-horsepower Anzani radials. A total of 2,450 G.3s were built, including a small number under license in Britain and Italy.

The Caudron G.4 was a larger, twin-engined version of the G.3, powered by two 80-horsepower Le Rhônes or 100-horsepower Anzanis. The Anzani-powered Caudron G.4s served mostly as training aircraft, and some, but not all, had their engines set up to turn in opposite directions to balance the torque of the whirling propellers. All the Le Rhône–powered Caudrons had both engines rotating in the same direction, clockwise from the pilot's orientation. Also, the two vertical tail surfaces of the G.3 were increased to four on the G.4.

The twin-engined configuration increased the range of the Caudron and provided a location for a forward-firing machine gun, typically a Hotchkiss or Lewis, although other types were also used. To protect against attacks from behind, some G.4s were fitted with an additional gun mounted on the top of the upper wing and pointed rearward, but this proved to be ineffective and was frequently removed from operational aircraft. A number of G.4s had a second gun mounted immediately in front of the pilot on the deck of the nacelle, as on the National Air and Space Museum's Caudron, but more often the pilot and observer simply carried handheld weapons to respond to attacks from the rear. Some G.4s carried a camera for high-altitude reconnaissance.

The prototype G.4 first flew in March 1915, and 1,358 were built in three major versions: the Caudron G.4A2 for reconnaissance, the G.4B2 for bombing, and the G.4E2 for training.

The A2 had a wireless set for artillery spotting missions; the B2 could carry up to 220 pounds of bombs; and the E2 had dual controls for instruction. A special armored version, designated the G.4IB—the "B" representing *blindage*, the French word for armor—was deployed to the top French units. In addition to reconnaissance, bombing, and training, the Caudron G.4 sometimes served as a long-range escort to other bomber aircraft.

By 1916, the G.4 was replacing the G.3 in most Caudron squadrons. Extensively used as a bomber during the first half of 1916, its deployment in that role was severely reduced by the fall of the same year. The Caudron's relative slow speed and inability to defend itself from the rear made it increasingly vulnerable to fighter attacks as German air defenses improved. But Caudrons continued to be widely used as reconnaissance aircraft well into 1917. By early 1918, however, virtually all Caudron aircraft still in use were relegated to training duties.

In addition to the French, British and Italian units used Caudrons extensively, and a few were used by the Russians and the Belgians. In November 1917, ten Caudron G.4s were sold to the United States and transferred to the US Air Service's 2nd Air Instruction Center at Tours. Used exclusively as trainers, none of these Caudrons saw operational combat service with American units.

The G.4 was in many respects a prewar design, with its wing-warping lateral control, light structure, and limited visibility. Yet it has great significance as an early light bomber and reconnaissance aircraft. It was a principal type used when these critical air-power missions were being conceived and pioneered in World War I, and although fighter aircraft frequently gain greater attention, the most influential role of aviation in World War I

Opposite: A Caudron G.4 and crew. The design was used for reconnaissance, bombing, and training.

Bottom left: The Smithsonian National Air and Space Museum Caudron G.4 was powered by two 80-horsepower Le Rhône rotary engines.

Bottom middle: The forward cockpit of the Caudron G.4. The spool and crank at the left lowered and raised a flexible antenna for the aircraft's wireless telegraph.

Bottom right: Essentially a pre–World War I design, the Caudron G.4 used the already antiquated wing-warping method for lateral control. When the G.4 was introduced into service in 1916, nearly all airplanes had movable surfaces called ailerons for roll control.

A stunning air-to-air photograph, rare for this period, shows a Caudron G.4 in flight.

was reconnaissance. The extensive deployment of the Caudron in this role makes it an especially important early military aircraft. Moreover, despite its speed and armament limitations, the Caudron was quite reliable, had a good rate of climb, and was pleasant to fly—all characteristics that made it a good training aircraft after its combat effectiveness was reduced. Many Allied pilots received their initial flight training on the Caudron.

The National Air and Space Museum's Caudron G.4, serial number C4263, also has great individual significance. It is among the oldest surviving bomber aircraft in the world and is one of the very few multi-engine aircraft from its period that remain. Especially significant, the airplane still carries its original fabric covering and paint, making it one of the most original World War I aircraft in existence. Only one other Caudron G.4 survives (in the Musée de l'Air et de l'Espace in Paris); however, that example has been completely restored.

C4263 was built by the Eugene firm and left the factory on December 12, 1916. The marking "12_16" appears on the leading edge of each individual wing section and some other major components of the aircraft, confirming the manufacturing date acquired from archival sources. Its acceptance flight was made at Issy-les-Moulineaux on December 27, 1916. The pilot's name was Gerviès and it was reported that the airplane climbed to 1,000 meters (3,281 feet) in seven minutes. The airplane had the full radio and photographic equipment that characterized the A2 reconnaissance variants.

The aircraft was not among the ten G.4s purchased late in 1917 and deployed as trainers and saw no operational activity with the French Air Service. C4263 was apparently sent to the Réserve Générale of the Aviation Militaire and remained there until it was purchased by the US government in early 1917 through the American ambassador, along with a Voisin Type 8 (also in the National Air and Space Museum collection) and a Farman aircraft, acquired for technical evaluation. However, by the time the aircraft were transported to the United States and prepared for flight demonstrations, they were already outmoded. The G.4 arrived in the United States and was at Langley Field, Hampton, Virginia, by July 26, 1917, four months before the contract for the ten training aircraft was executed.

On July 12, 1918, Lt. Col. L. S. Horner of the War Department's Bureau of Aircraft Production wrote to Smithsonian Institution secretary Charles Walcott regarding "obsolete airplanes for exhibition purposes," offering the Caudron, the Voisin, and the Farman to the Smithsonian. The offer was accepted, and the three airplanes were delivered on September 16 and 17, 1918. The Farman was very incomplete and was deemed unacceptable for exhibition. It was returned to the War Department in June 1921. Because of an oversight when packing the Farman for shipment, however, its wings remained at the Smithsonian until September 1925, when they were either returned to the War Department or destroyed.

The Caudron was delivered to the Smithsonian without engines, propellers, or armament but was soon assembled and displayed without these components. In 1929, engines and propellers were acquired from the War Department and installed. Unfortunately, the museum was unable to obtain the proper 80-horsepower Le Rhône rotary engines. Only 110-horsepower Le Rhônes were available and were considered at least to be representative of the correct powerplants. Similarly, the propellers acquired and installed on the airplane were only representative of the period and not the precise type that were actually used on the Caudron. Internal parts of the 110-horsepower Le Rhônes, such as pistons and connecting rods, were removed to lighten the engines and put in storage. Smithsonian officials took this action as a safety measure because the airplane was suspended over a public area. The Caudron was displayed in this manner until the late 1960s or early 1970s, when it was placed in storage. It received preservation treatment in 2000 in preparation for display at the museum's soon-to-open Steven F. Udvar-Hazy Center. All components were carefully conserved to retain the originality of this exceedingly rare specimen. At that time the 110-horsepower Le Rhône engines were replaced with the proper 80-horsepower Le Rhônes, and one original correct propeller was acquired and a second fabricated using the original as a pattern.

One interesting marking on the tail of the National Air and Space Museum Caudron G.4 is the term "Blindage 16K" on the rudders, indicating the Museum's airplane could be one of the special G.4IB armored Caudrons. The armor consisted of a heavy metal plate inserted behind the seat of the rear cockpit, protruding upward so as to cover the back and head of the pilot. The armor plate is missing on the National Air and Space Museum Caudron; however, there is a gap, or slot, in the structure immediately behind the seat where such a plate would fit. Photographs of other Caudrons with the armor plate inserted show it in the same location as the slot behind the rear seat on the Museum's Caudron. This, along with the "Blindage 16 K" marking, strongly indicates that the airplane was an armored version of the Caudron G.4.

SPECIFICATIONS
CAUDRON G.4

WINGSPAN: 56 ft. 5 in. (17.2 m)

23 ft. 8 in. (7.2 m)

8 ft. 6 in. (2.6 m)

EMPTY WEIGHT: 1,616 lb. (733 kg)
GROSS WEIGHT: 2,716 lb. (1,232 kg)

ENGINES: 2 x Le Rhône 9C rotary
9-cylinder, 80 hp each
MAXIMUM SPEED: 77 mph (124 km/h)

DOUGLAS WORLD CRUISER DWC-2 *CHICAGO*

CHAPTER 5

BY ALEX M SPENCER

With the successful crossings of the Atlantic in 1919 by the US Navy's NC-4 and the British flyers John Alcock and Arthur Brown in a Vickers Vimy, the ambition to circumnavigate the globe by airplane was a natural next challenge. The British made an unsuccessful attempt in 1922. The following year a French team also tried and failed, while another British effort was begun but abandoned. Italian and Portuguese teams were also discussing plans for a round-the-world flight.

In July 1923, the US War Department disclosed that it was sending two officers on an information-gathering trip to stake out a route for a global flight to be attempted by the US Army Air Service in 1924. Major General Mason M. Patrick, chief of the air service, was put in charge of planning and directing the flight. The publicly stated objectives of the world flight were to establish air routes, improve commerce, and foster better international relations, but its underlying intent was to garner popular support for the army air service and the creation of an independent air force. Any benefits relating to the creation and expansion of commercial air routes were slow to materialize.

Japanese mechanics inspect the engine of one of the World Cruisers during one of the round-the-world aerial expedition's six stops in Japan.

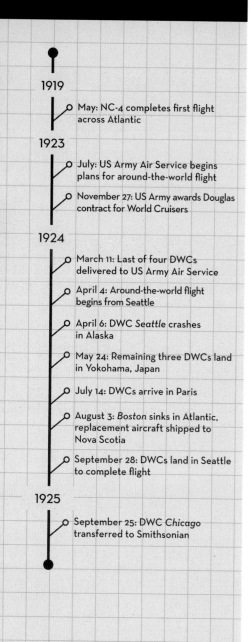

TIMELINE
DOUGLAS WORLD CRUISER DWC-2 *CHICAGO*

1919

May: NC-4 completes first flight across Atlantic

1923

July: US Army Air Service begins plans for around-the-world flight

November 27: US Army awards Douglas contract for World Cruisers

1924

March 11: Last of four DWCs delivered to US Army Air Service

April 4: Around-the-world flight begins from Seattle

April 6: DWC *Seattle* crashes in Alaska

May 24: Remaining three DWCs land in Yokohama, Japan

July 14: DWCs arrive in Paris

August 3: *Boston* sinks in Atlantic, replacement aircraft shipped to Nova Scotia

September 28: DWCs land in Seattle to complete flight

1925

September 25: DWC *Chicago* transferred to Smithsonian

For the flight, the air service commissioned specially built aircraft from the Douglas Aircraft Company in Santa Monica, California, requiring a strong airplane that could endure harsh conditions ranging from arctic cold to tropical heat and humidity. With this in mind, the army planners chose a modified navy Douglas DT-2 torpedo bomber, a sturdy, two-place biplane powered by a single 12-cylinder, water-cooled, 400-horsepower Liberty engine. Because the flight would take place over both land and water, the airplane had to have interchangeable floats and wheels.

To increase range, fuel tanks were added to the upper wing center section, to the lower wing roots, behind the firewall, and under the pilot's seat. These tanks increased the fuel capacity from 115 gallons to 644 gallons. Other modifications included a larger radiator, strengthened bracing, increased rudder surface, a cutout in the upper wing for increased visibility, and the relocation of the observer's cockpit closer to the pilot's. A total of five airplanes were built, including a prototype for testing. The Douglas World Cruisers, as they were named, did not have radios or advanced navigational aids, only the standard rudimentary flight instrumentation of the day.

The four World Cruisers built for the round-the-world attempt were christened the *Seattle*, the *Chicago*, the *Boston*, and the *New Orleans* to represent the regions of the United States. Four of the air service's top pilots were selected: Major Frederick L. Martin was designated overall flight commander and flew the *Seattle*; Lts. Lowell H. Smith, Leigh Wade, and Erik H. Nelson piloted the *Chicago*, the *Boston*, and the *New Orleans*, respectively. Each pilot was permitted to select his own mechanic/copilot. The crews trained in navigation and meteorology at Langley Field, Virginia, practicing with the prototype airplane while the flight aircraft were being constructed and prepared.

In the meantime, elaborate preparations were made. Securing landing rights for the world flight became a diplomat's bad dream, as the flight would touch twenty-eight different nations and colonial mandates. Only six years following the end of World War I, international mistrust, civil wars, and the threat of revolutions all along the route provided hazards beyond weather and geography. The US State Department faced difficult negotiations to secure overflight and landing clearances. For example, because the United States did not recognize the Soviet Union at the time, flying over Siberia was prohibited, necessitating a Southeast Asian route that added 6,875 miles to the journey. Also, the Japanese refused to allow the flight to land until the United States promised that crews would not take photographs of the home islands. Since the vast majority of the flights in Asia occurred over British colonial possessions, the services of the Royal Air Force were secured.

Beyond these diplomatic problems, the preparations for the world flight were a massive logistical undertaking, and detailed preparation was critical to its success. Flight planners arranged to make seventy-four stops and cover about 27,550 miles. Thousands of gallons of fuel and oil, thirty-five replacement engines, and numerous spare parts for the airplanes had to be distributed throughout the world, some in places where airplanes had never flown before. Each leg of the flight required over 1,400 gallons of gasoline to fuel the four planes. The task of establishing these remote supply depots fell to the US Navy and Coast Guard. Planners divided the world and flight into six geographic regions to ease the organization of equipment and supplies.

After practice with the four aircraft in Santa Monica and San Diego, the crews headed north to Seattle, the official point of departure, on April 6, 1924. On the first phase of the flight, the World Cruisers crossed the North Pacific, a feat never before accomplished by

The *Chicago* on display in the Baron Hilton Pioneers of Flight Gallery.

airplanes. In many ways the first legs of the journey were the most difficult—the flyers faced freezing temperatures, thick and unpredictable fog, and sudden, violent storms.

Shortly after setting off, Major Martin, piloting the lead airplane, fell behind; mechanical problems had delayed the takeoff of the *Seattle* from Kanatak, Alaska. After repairing the airplane, Major Martin and copilot Sgt. Alva Harvey continued on toward Chignik, Alaska, to catch up to the other World Cruisers. While flying over uncharted and unfamiliar terrain, however, they encountered a rapidly developing storm. Blinded by a thick cloud cover and unable to pinpoint any landmarks, Martin crashed the airplane into a mountainside. Fortunately, neither man was injured. For ten days, equipped only with their emergency ration kit and service pistols, Martin and Harvey hiked through the Alaskan wilderness. They finally came upon a remote fishing village at Port Moller and were rescued. Lt. Lowell Smith, pilot of the *Chicago*, assumed command for the remainder of the flight.

The crews of the Douglas World Cruisers (from left): Sgt. Arthur Turner, Sgt. Henry Ogden, Lt. Leslie Arnold, Lt. Leigh Wade, Lt. Lowell Smith, Maj. Frederick Martin, and Sgt. Alva Harvey

Unable to fly over Siberia, the expedition was faced with Japanese authorities who feared that the United States would use the flight to map a potential air-invasion route from the Aleutian Islands. After intense negotiations, officials reluctantly allowed a serpentine route that prevented the airplanes from flying over any military bases. At numerous locations throughout Japan, large crowds greeted the Americans, but suspicious authorities did not allow the American pilots to go ashore.

Continuing on to the coast of China, the flyers encountered a nation rocked by political turmoil. The country was split by civil war, and many provinces were ruled by warlords or were under European colonial rule. During this stage of the flight, the flyers also faced typhoons, disease, and extreme heat and humidity. The environmental conditions placed great stresses on the engines and airframes, and the *Chicago*, in particular, was hampered by constant mechanical failures.

When the flyers arrived in Rangoon, Burma, they intended to replace the airplanes' floats with wheels, but the early arrival of monsoons threatened to cancel the flight entirely. To avoid any further delays, Lieutenant Smith decided to postpone the scheduled replacement of the floats as well as the airplane engines and press forward.

Continuing on, the pilots and their airplanes faced extreme heat and blowing sands over the Middle Eastern deserts. Armed nationalist uprisings in Mesopotamia (present-day Iraq)

added to the danger. As the World Cruisers crossed Europe, the trip passed with relative good fortune. They still had to traverse numerous national borders where no international agreements existed for flights. From Strasbourg, France, they were escorted to Paris by the French air force and there received a tumultuous welcome from cheering crowds on July 14, Bastille Day. The airmen lunched there with Gen. John Pershing, the commander of the American Expeditionary Force during World War I, and the following day left Paris and landed in London. The only real delays that the airmen faced in Europe were the result of the many receptions given to them.

With the daunting Atlantic crossing ahead, the World Cruisers were re-equipped with floats at Brough, England. Over the ocean, the pilots faced the journey's longest overwater flights and had no alternative landing sites. The US Navy placed a series of picket ships along the route to rescue the pilots if they had to land in the open ocean. Dense fog and sudden storms were a continual problem, and at one point, low cloud cover forced the pilots so close to the water that they had to dodge the peaks of icebergs.

Disaster occurred again over the North Atlantic between the Orkney and Faroe islands. On the first Atlantic leg, from Kirkwall in the Orkneys to Hornafjörður, Iceland, the *Boston* experienced trouble with its oil pump and was forced to alight at sea. For almost four hours, its crew, Lieutenants Wade and Ogden, awaited rescue. They were found and their airplane

This aerial photograph depicts the *Chicago* in flight and gives a sense of the tight formation these aircraft were flying at that moment.

A front view of the Douglas World Cruiser *Chicago* on display in the Baron Hilton Pioneers of Flight Gallery.

taken in tow by a British trawler, the *Rugby-Ramsey*. An hour later the USS *Richmond* took over the towing.

The seas and wind began to rise and threatened to overwhelm the airplane. Crews tried to lift the *Boston* onto the *Richmond*'s deck, but a large wave rolled the ship, causing its boom and tackle to strike and severely damage the World Cruiser. After an attempt to save the *Boston*, the airplane was cut loose and sank.

In Reykjavik, the crews of the *Chicago* and *New Orleans* learned of the *Boston*'s fate. Once final preparations were completed for the longest overflight of trip, the two remaining aircraft made their way to Greenland. After the long and harrowing flight across the Atlantic, the World Flight finally arrived back in the United States for the final stages of the journey. When the aircraft arrived over New York City, *New Orleans* and *Chicago* were joined by the *Boston II*, which would accompany them for the rest of the trip. Large and enthusiastic crowds greeted them at each landing site across the country. To avoid crossing the dangerous Rocky Mountains, the pilots flew through the desert Southwest and then across southern California and up the West Coast to Seattle. On September 28, 1924, the World Cruisers made their final landing at Sand Point, the same field from which they had departed 175 days earlier. A reporter asked Lt. Lowell Smith if he would be willing to make the flight again. He replied, "Not in a million years, unless ordered to do so."

The incredibly arduous 1924 round-the-world flight remains one of the truly great achievements in aviation. The loss of two of the airplanes and the close call for Major Martin and Sergeant Harvey in the crash of the *Seattle* were hardly the only setbacks. Throughout the journey the crews prevailed against an endless series of forced landings, repairs, bad weather, and other mishaps that continually threatened their success. Further, it was a monumental logistical accomplishment. More than just an aviation milestone, the flight was an important step toward the goal of worldwide air transport.

Two weeks before the Douglas World Cruisers completed their flight around the world, a young museum aide named Paul E. Garber recommended that the Smithsonian Institution acquire one of the aircraft for its collection. Eleven months later, the Secretary of War approved the transfer of the *Chicago* to the Smithsonian. On September 25, 1925, this aircraft made its final flight from McCook Field in Dayton, Ohio, to Bolling Field in Washington, D.C. Later that fall, the airplane was placed on public display in the Smithsonian's Arts and Industries building.

The *Chicago* was restored between 1971 and 1974, and it was moved into the new National Air and Space Museum building in 1976. Of the five Douglas World Cruisers built, the *New Orleans* is the only other survivor and is in the collection of the Los Angeles County Museum of Natural History.

The original World Flight fabric insignia was removed from the *Chicago* during its restoration from 1971 to 1974 at the Smithsonian National Air and Space Museum.

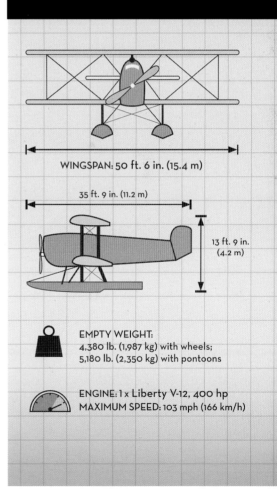

SPECIFICATIONS
DOUGLAS WORLD CRUISER DWC-2 *CHICAGO*

WINGSPAN: 50 ft. 6 in. (15.4 m)

35 ft. 9 in. (11.2 m)

13 ft. 9 in. (4.2 m)

EMPTY WEIGHT:
4,380 lb. (1,987 kg) with wheels;
5,180 lb. (2,350 kg) with pontoons

ENGINE: 1 x Liberty V-12, 400 hp
MAXIMUM SPEED: 103 mph (166 km/h)

RYAN NY-P
SPIRIT OF ST. LOUIS

CHAPTER 6

BY F. ROBERT VAN DER LINDEN

"Our messenger of peace and goodwill has broken down another barrier of time and space." So spoke President Calvin Coolidge about Charles A. Lindbergh's extraordinary solo transatlantic flight in 1927. Not until the *Apollo 11* moon landing in 1969 would the entire world again be as enthusiastic about an aviation event as it was when Lindbergh landed his little Ryan monoplane in Paris.

Lindbergh was an obscure twenty-five-year-old airmail pilot when he made his epic flight. The son of a Minnesota congressman and a teacher, Lindbergh, though highly intelligent, had little interest in formal education, preferring to toy with mechanical devices while splitting time between running the family farm in Little Falls, Minnesota, and visiting his father in Washington, D.C.

In 1922, after a less-than-distinguished year and a half at the University of Wisconsin, Lindbergh left to learn to fly with the Nebraska Aircraft Corporation. While the lessons were sketchy, he found his passion and began performing as a barnstormer until 1924, when he enrolled as a flying cadet in the army air service, an experience that transformed him. He became highly disciplined and detail oriented—traits that would serve him well as an aviator. He also learned the value of an education and excelled in his training. He won his reserve commission and began serving as a civilian

Flags of the countries Lindbergh and the *Spirit* visited after his historic flight were painted on the nose. The cowling's light gold color is a coating of varnish that was applied to protect the aluminum alloy from corrosion.

1919

May 22: Raymond Orteig offers $25,000 prize to first aviator to fly nonstop from New York City to Paris

1922

Charles Lindbergh learns to fly

1924

Lindbergh becomes flying cadet in US Army Air Service

1927

February: Donald Hall begins design of Ryan NYP

April: Ryan Aircraft completes construction of NYP *Spirit of St. Louis*

May 10–12: Lindbergh sets cross-continental flight record in *Spirit of St. Louis*

May 20–21: Lindbergh completes solo Atlantic crossing from Roosevelt Field, New York, to Le Bourget Field, Paris

June 11: Lindbergh and *Spirit of St. Louis* return from Europe on USS *Memphis*

July–September: Lindbergh tours United States with *Spirit of St. Louis*

1927–1928

December 1927–February 1928: Lindbergh tours Latin America and Caribbean in *Spirit of St. Louis*

1928

April 30: *Spirit of St. Louis* arrives at Smithsonian

airmail pilot, joining the Robertson Aircraft Corporation in April 1926 to fly the Contract Air Mail Route 2 between St. Louis and Chicago.

That September, Lindbergh learned of the $25,000 prize offered by New York hotel owner and French ex-patriot Raymond Orteig for the first successful nonstop flight between New York and Paris. Inspired, Lindbergh sought financial backing from the Robertson brothers and several St. Louis businessmen who agreed to help. With their support, Lindbergh next traveled to New York City to purchase an aircraft but was rebuffed. The aircraft he wanted, the sole existing Wright-Bellanca WB-2, was not available, and while its designer, Giuseppe Bellanca, was willing to build him another

aircraft, Bellanca reserved the right to choose the pilot. This was unacceptable to Lindbergh. The Fokker aircraft company offered a $90,000 trimotor, but Lindbergh had already rejected the idea of a large multiengine aircraft as too complex, having learned of the fatal accident of René Fonck's Sikorsky S-35 the previous September.

Lindbergh wanted a small, single-engine monoplane of conventional, proven design with which he could trust his life. After Travel Air of Wichita turned him down, Lindbergh approached Ryan Airlines, and the small company located in San Diego, California, agreed to build him an aircraft. Development began based on a standard Ryan M-2, with Donald A. Hall as principal designer. Because of the nature of the flight, modifications had to be made to the basic high-wing, strut-braced monoplane design. The wingspan was increased by 10 feet, and the structural members of the fuselage and wing cellule were redesigned to accommodate the greater fuel load. Plywood was fitted along the leading edge of the wings. The fuselage design followed that of a standard M-2 except that it was lengthened 2 feet. The cockpit was moved farther to the rear for safety and the engine moved forward for balance, permitting the fuel tank to be installed at the center of gravity. The aircraft featured a welded steel tube fuselage and wooden wing, all covered in grade A cotton fabric and sealed with a thin coat of aluminum-colored dope.

For Lindbergh, fuel was life—as long as the engine ran and he could head east, he assumed that he would eventually find land, but he needed as much gasoline as he could carry. The main tank was installed behind the instrument panel, which took up the space normally occupied by the windscreen. Lindbergh could see forward only by means of a periscope or by turning the aircraft to look out of a side window. A second fuel tank was placed in the nose behind the 25-gallon oil tank, while three gravity-fed fuel tanks were installed in the wings. A clever system of tubing and petcocks allowed Lindbergh to select which tank to use in order to burn off the fuel evenly. Though the aircraft was designed to carry 425 gallons, Lindbergh was able to squeeze 450 gallons into it for his transatlantic flight.

A revolutionary Wright Whirlwind J-5C supplied the power. This remarkable engine was a recent development and was used by all of the transatlantic contestants that spring. What made the engine so remarkable was the fitting of spring-loaded grease cups that provided a constant supply of grease during the long flight. Of even greater significance was the fact that the J-5C was the first engine fitted with sodium-cooled exhaust valves. With the hollow stems of the valves filled partway with pure sodium, the heat from the engine's combustion was efficiently drawn away from the cylinder and into the cooling stream of air around the engine. This clever invention of British engineer Samuel Heron virtually cured the incessant problem of burned exhaust valves in early aircraft engines. Thus, Lindbergh's airplane had a powerful, lightweight, and reliable powerplant. Ironically, because Lindbergh was a little-known airmail pilot, Wright

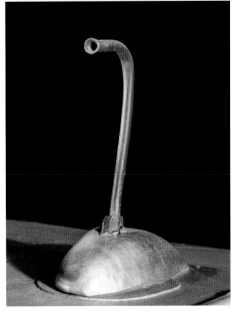

Opposite: Lindbergh works on the engine just prior to his Paris flight.

Top: Lindbergh used these petcocks below the instrument panel to control fuel from the *Spirit*'s five tanks, thus burning the 450 gallons evenly.

Bottom left: Lindbergh set the aircraft's Earth inductor compass using this instrument on the right side of his seat.

Bottom right: This detail photograph shows an air vent for one of the three fuel tanks located in the wing.

assigned its newest engine builder to Lindbergh's J-5C. Luckily, mechanic Tom Rutledge knew his craft.

In only two months, Ryan designer Donald Hall and his small team planned and built the unique aircraft, designated the NYP, with significant input from Lindbergh, who insisted that the metal fuel lines be cut every 18 inches and secured with a rubber fitting to absorb vibration. Lindbergh had learned this trick while flying aging de Havilland DH.4 mail planes that had a tendency toward cracked fuel lines. At one point Hall pleaded with Lindbergh to allow him to fit a larger fin and rudder for better stability, but Lindbergh wisely refused. He was concerned that the redesign might take too long, and the aircraft had to be ready by May when the skies over the North Atlantic cleared. Of greater significance, Lindbergh understood that an unstable aircraft would force him to stay awake, because it would turn suddenly if he fell asleep and his hand fell off the control stick. This feature would save his life when he caught himself dozing several times on the flight.

Work on the aircraft was completed late in April 1927. It was painted and carried registration number N-X-211 painted in black. The aircraft was christened the *Spirit of St. Louis* at the suggestion of Harold Bixby, one of Lindbergh's backers, in honor of the pilot's friends and associates in Missouri who financed the flight.

After several test flights, Lindbergh flew the aircraft from San Diego to New York on May 10, 11, and 12, making only one stop—at St. Louis. His cross-country flight time of twenty-one hours and forty minutes set a new transcontinental record. During this flight, his engine almost failed while crossing the southern Rocky Mountains, and as a result he had a carburetor heater installed to prevent engine icing despite the added weight. This, too, would prove effective when he crossed the cold, damp Atlantic. When Lindbergh arrived in New York, an attentive mechanic noticed that the propeller spinner, in which all of the Ryan factory employees had their names painted, had developed a serious crack. On his own initiative, the mechanic designed and hand-built a replacement that remains on the aircraft to this day.

After waiting several days in New York for favorable weather, Lindbergh took off for Paris on the morning of May 20, 1927. He had not intended to fly that day, but the previous evening he learned that there was an expected break in the cloud cover over the ocean, so he stepped up his preparations despite getting no sleep. His lengthy takeoff with a slight tailwind along a muddy airfield in his overweight airplane led to several dramatic moments, none more so then when he barely cleared the telephone lines at the end of the field. Lindbergh and the *Spirit* struggled for altitude as he flew up the coast of New England, but gradually, as the fuel burned off, he gained a safe altitude for the rest of the flight.

Over the next thirty-three hours, thirty minutes, and 3,610 miles, Lindbergh fought exhaustion. Despite the challenges and his inexperience navigating over water, he made landfall in Ireland within 3 miles of his target and then flew on to France. Finding Paris, the City of Lights, was easy, but he did not know exactly where to find Le Bourget Airport. Taking an educated guess, Lindbergh followed a line of lights stretching out from the city to a large field. Unknowingly, he was following the headlights of the massive traffic jam pouring to the airport to meet him. He landed safely at Le Bourget Field, where he was greeted by a wildly enthusiastic crowd of one hundred thousand.

Lindbergh was not expecting this incredible reception and had, in fact, brought letters of introduction just in case he got into trouble with the authorities. He was carried to safety by French air force officers and taken under wing by US Ambassador Myron Herrick. The *Spirit* was not so lucky. Thousands of well-wishers and souvenir hunters pressed against the aircraft, tearing off pieces of fabric before authorities regained control. It was quickly towed to safety, but not before holes had been torn in the fuselage and ailerons. The French air force was so embarrassed that they replaced all of the fuselage fabric with linen and made

 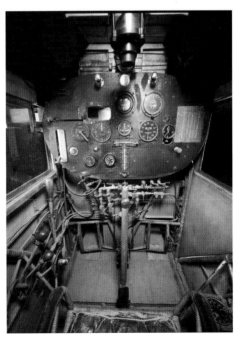

Left: Lindbergh sat in this wicker seat for 33½ hours during his epic flight to Paris.

Right: Lindbergh's view in front of his aircraft was blocked by two fuel tanks and the instrument panel. He used a periscope on the left of the panel as well the side windows to see ahead.

spot repairs elsewhere overnight. By the time Lindbergh woke up two days later, the *Spirit of St. Louis* looked as good as new.

Lindbergh and his aircraft returned to the United States aboard the light cruiser USS *Memphis*—sent by President Calvin Coolidge to make sure the most famous man in the world returned safely to America—to a hero's welcome on June 11. Lindbergh received tumultuous welcomes in Washington, D.C., and in New York City. From July 20 until October 23 of that year he took the famous plane on a tour of the United States. It has been estimated that more than one-third of the country's population came out to see him on this tour. Then, on December 13, he and the *Spirit* flew nonstop from Washington, D.C., to Mexico City; through Central America, Colombia, Venezuela, and Puerto Rico; and nonstop from Havana to St. Louis. Beginning in Mexico City, flags of the countries he visited were painted on both sides of the cowling.

Lindbergh did much more than fly across the Atlantic. His fame enabled him to achieve a great deal in promoting air travel and science. His work with Transcontinental Air Transport, Trans World Airlines (TWA), and Pan American Airways was instrumental in the success of these pioneering airlines. Through the Guggenheim Foundation for the Promotion of Aeronautics, Lindbergh helped to sponsor the research of rocket pioneer Robert Goddard, and working with French surgeon Dr. Alexis Carrel he helped to develop a practical perfusion pump that was a step toward the development of an artificial heart. Lindbergh's celebrity, however, also came at great personal and professional cost. Hounded by the media, his privacy lost, he and his wife, Anne Morrow Lindbergh, suffered the kidnapping and murder of their first child in 1932. In his later life, Lindbergh's polarizing personal and political beliefs came under increasing dislike by the American public, adding layers of complexity to this American hero.

The Smithsonian Institution is a part of the Lindbergh story. When Charles Lindbergh awoke in Paris that morning in 1927, included among the countless congratulatory notes was a telegram from Smithsonian secretary Charles G. Abbott requesting the *Spirit of St. Louis* for the National Collection. Legendary curator Paul E. Garber wrote the request for the secretary, and the aviator and his St. Louis backers eagerly agreed. Following the completion of Lindbergh's successful US and Latin American tour, he sold the *Spirit* to the Smithsonian—for $1.

On April 30, 1928, the aircraft arrived in Washington, and two weeks later it was installed in the Arts and Industries Museum.

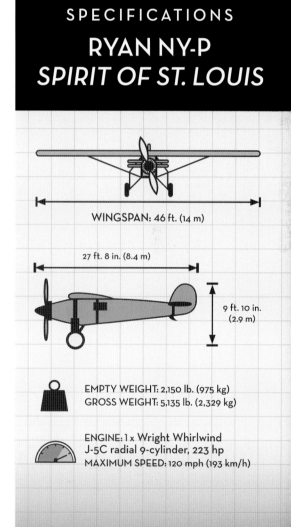

SPECIFICATIONS
RYAN NY-P
SPIRIT OF ST. LOUIS

WINGSPAN: 46 ft. (14 m)

27 ft. 8 in. (8.4 m)

9 ft. 10 in. (2.9 m)

EMPTY WEIGHT: 2,150 lb. (975 kg)
GROSS WEIGHT: 5,135 lb. (2,329 kg)

ENGINE: 1 x Wright Whirlwind J-5C radial 9-cylinder, 223 hp
MAXIMUM SPEED: 120 mph (193 km/h)

LOCKHEED VEGA 5B AND LOCKHEED VEGA 5C
WINNIE MAE

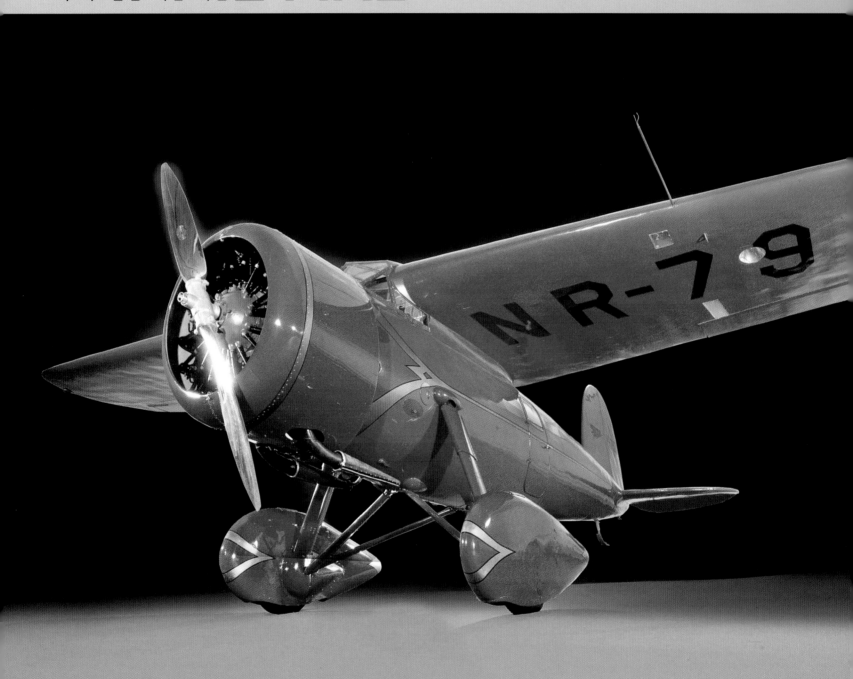

CHAPTER 7

BY DOMINICK A. PISANO

Introduced in 1927, the Vega, designed by John Northrop and Gerard Vultee for Lockheed Aircraft, was undoubtedly the most modern aircraft of its time. Although it was constructed of wood, the Vega had an enclosed cockpit, internally braced (cantilever) wing, and molded shell (monocoque) construction in which the outer skin of the fuselage carried a major part of the stresses placed on the plane. It was also one of the first truly streamlined aircraft, designed specifically to fly swiftly and efficiently through the air with the least amount of resistance, and the first production aircraft to carry the drag-reducing National Advisory Committee for Aeronautics (NACA) cowling around its Pratt & Whitney radial engine.

Two Vegas, both residing in the National Air and Space Museum Collection, have the distinction of being among the United States' most historically significant, record-setting aircraft: the Vega 5B flown by Amelia Earhart and the Vega 5C piloted by Wiley Post. These aviators' record-setting flights at the controls of their respective aircraft were made during the so-called Golden Age of Aviation, during which technological, political, financial, and sociocultural developments helped make the airplane a symbol of American ingenuity.

The Lockheed Vega flown by Amelia Earhart is on display in the National Air and Space Museum's Barron Hilton Pioneers of Flight Gallery.

1927
○ Vega first flies

1929
○ August: Amelia Earhart places third in Vega in Women's Air Derby Race

1930
○ Wiley Post sets transcontinental speed record during National Air Races

1931
○ June 23–July 1: Post and Harold Gatty complete eight-day around-the-world flight in *Winnie Mae*

1932
○ May 20–21: Earhart flies Vega 5B to become first woman to pilot aircraft across Atlantic

○ August 24–25: Earhart completes first solo nonstop transcontinental flight by woman

1933
○ Earhart sells Vega to Franklin Institute

○ July 15– 22: Post completes first solo around-the-world flight in *Winnie Mae*

1935
○ March 15: Post uses first pressure suit to fly in stratosphere and sets speed record from Los Angeles to Cleveland

○ August 15: Post and Will Rogers die in Alaskan crash of modified Lockheed Explorer

1936
○ June 30: Smithsonian acquires Lockheed 5C *Winnie Mae* from Post's widow

1937
○ July 2: Amelia Earhart disappears in Pacific in Lockheed Electra

1966
○ Smithsonian acquires Earhart's Vega from Franklin Institute

Both of Earhart's historic flights in the Vega 5B (registration number NC-7952) were made in 1932: the first solo nonstop flight across the Atlantic Ocean by a woman and the first solo nonstop flight across the United States by a woman. The aircraft seemed to be made for Earhart, and she owned a number of them during her flying career. One of the world's most celebrated woman pilots, she was born on July 24, 1897, in Atchison, Kansas, and graduated from Hyde Park High School in Chicago, where she excelled in science, before enrolling in the exclusive Ogontz School outside Philadelphia. She did not graduate, however, choosing to volunteer at Toronto's Military Hospital to care for soldiers wounded in World War I. It was in Toronto that that she attended a flying exhibition in which a low-flying stunt pilot passed right above her and a friend. Earhart, who did not budge, later wrote, "I did not understand it at the time, but I believe that little red airplane said something to me as it swished by."

In 1919 and 1920, Earhart briefly attended Columbia University as a pre-med student but eventually left to join her parents who, after a long separation, were living in Los Angeles. In 1921 she took her first flying lesson from an instructor named Neta Snook. Working at various jobs, she managed to save enough money to continue her lessons and purchase a yellow Kinner Airster, which she named the *Canary*. On December 15 of that year she received her pilot's license.

After a somewhat long hiatus from flying, Earhart gained celebrity in 1929, but not as a pilot. A woman named Amy Phipps Guest, the daughter of American industrialist and

philanthropist Henry Phipps, who was the partner of Andrew Carnegie at Carnegie Steel Company, wanted to fly across the Atlantic. Guest owned the Fokker F.VII *Friendship*, in which she intended to make the passage, but her family objected. She asked famed arctic explorer Richard E. Byrd and New York publisher George P. Putnam to select "another girl with the right image" for the flight.

Earhart was interviewed and selected. However, she was merely a passenger on the *Friendship*, which was piloted on a twenty-hour, forty-minute flight from Newfoundland to Wales by Wilmer Stultz and Lou Gordon. Earhart later remarked that she felt like "a sack of potatoes" during the flight. Nevertheless, the group was celebrated with a ticker-tape parade in New York City and a White House reception with President Calvin Coolidge. The *Friendship* flight was just the beginning of Earhart's celebrity.

In August 1929, she flew a Lockheed Vega, registration number NC-31E, in the Women's Air Derby Race, an offshoot of the prestigious National Air Races dubbed the "Powder Puff Derby" by humorist Will Rogers. The race was to be flown from Santa Monica, California, to Cleveland, Ohio. Entry requirements were rigid and the same as those for men who competed in the National Air Races—contestants had to have one hundred hours' solo flying time and to have flown twenty-five hours cross-country. Among the twenty pilots who qualified for the race were a number of women who would become notable in flying circles, including Louise M. Thaden, Blanche W. Noyes, Florence "Pancho" Barnes, and Ruth Elder.

Earhart did not win, placing third, but in 1929 she and the winner, Thaden, went on to found the Ninety-Nines, the first organization "to promote women pilots among themselves, and to encourage other women to fly, as well as to break down general opposition to aviation." Earhart became the first president of the organization.

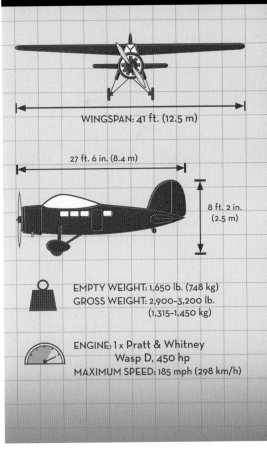

SPECIFICATIONS
LOCKHEED VEGA 5B

WINGSPAN: 41 ft. (12.5 m)

27 ft. 6 in. (8.4 m)

8 ft. 2 in. (2.5 m)

EMPTY WEIGHT: 1,650 lb. (748 kg)
GROSS WEIGHT: 2,900–3,200 lb. (1,315–1,450 kg)

ENGINE: 1 x Pratt & Whitney Wasp D, 450 hp
MAXIMUM SPEED: 185 mph (298 km/h)

An interior view of Amelia Earhart's Lockheed 5B Vega shows the instrument panel and control stick.

On May 20–21, 1932, exactly five years after Charles Lindbergh's historic solo transatlantic flight in the *Spirit of St. Louis*, Earhart flew her Vega in a fifteen-hour, 2,026-mile nonstop flight from Harbor Grace, Newfoundland, to a spot near Londonderry, Northern Ireland, becoming the first woman to cross the Atlantic solo. For her accomplishment, the US Congress awarded her the Distinguished Flying Cross, and the French government awarded her the Cross of Knight of the Legion of Honor. In addition, President Herbert Hoover presented her with the Gold Medal of the National Geographic Society, and on June 20, 1932, New York City honored her with another ticker-tape parade.

A scant three months later, on August 24–25, 1932, Earhart took the Vega on another record-setting flight, this time across the United States, becoming the first woman to fly solo and nonstop coast to coast and setting a women's nonstop transcontinental speed record, flying 2,447.8 miles in nineteen hours, five minutes.

After purchasing a Vega 5C, Earhart sold her 5B to the Franklin Institute in 1933. The Smithsonian Institution acquired it in 1966.

In July 1937, Earhart and her navigator, Fred Noonan, disappeared in the vicinity of Howland Island, north of the Equator in the Central Pacific Ocean, during an around-the-world flight attempt in a Lockheed Electra. Her disappearance has generated intense interest to this day, spawning numerous conspiracy theories as well as legitimate attempts to discover the mysterious circumstances that surround her final act.

The names *Winnie Mae* and Wiley Post are synonymous with historic around-the-world flights, the development of the world's first functional pressure suit, and early attempts at substratospheric flight.

The Smithsonian National Air and Space Museum's Lockheed 5C Vega *Winnie Mae* is on display in the Time and Navigation Gallery.

Born on November 22, 1898, near Grand Saline, Texas, Wiley Hardeman Post—the fourth of four sons of William Francis Post and Mae Quinlan Post—grew up in Oklahoma farm country. Inspired by an aerial demonstration given by early exhibition pilot Art Smith in 1913, Post left home soon after, using money he had earned picking cotton to enroll in a seven-month course at the Sweeney Auto School in Kansas City. During World War I, Post joined the Students Army Training Corps in Norman, Oklahoma, and took courses in radio. After the war ended, he found work in the Oklahoma oilfields. In 1924, when barnstorming company Burrell Tibbs and his Texas Topnotch Fliers came to the area, Post cajoled Tibbs into letting him do parachute jumps as part of the show. Tibbs eventually relented.

Post later began parachute jumping on his own, making ninety-nine jumps over a period of two years. In the meantime, he had slowly been learning to fly, and in 1926 he made his first solo flight. Eager to earn enough money to buy his own airplane, Post went back to working in the oil fields. An accident in which a metal chip lodged in his left eye led to the eye's removal, the reason he is often pictured wearing an eye patch. Despite this misfortune, he trained his good eye to compensate for the vision loss. Post became the personal pilot for Oklahoma oilmen Powell Briscoe and F. C. Hall, flying them on various business trips in a Travel Air biplane, for which he was paid about $200 a month.

A National Air and Space Museum mannequin wears Post's high-altitude suit.

Later, Hall was so impressed by the performance of the Lockheed Vega when he was given a demonstration flight that he bought one and named it after his daughter, Winnie Mae. Circumstances brought on by the Great Depression forced Hall to sell Winnie Mae (NC 7954) back to Lockheed, which in turn sold it to Nevada Airlines. The chief pilot and operations manager for Nevada Airlines was the celebrated racing pilot Roscoe Turner, who, in August 1929, used the aircraft—renamed Sirius and carrying four passengers, including the most experienced aerial navigator in the United States, an Australian named Harold Gatty—to set a transcontinental speed record of nineteen hours, fifty-three minutes. Turner also flew this Vega in the 1929 National Air Races, placing third.

By 1930, Hall's circumstances had improved, and in June he purchased another Lockheed Vega (NC 105-W), which he also named Winnie Mae. The first significant flight Post made in the new Winnie Mae was as an entrant in the 1930 National Air Races' Men's Non-Stop Derby. Post flew the Vega from Los Angeles to Chicago in nine hours, nine minutes, four seconds, earning a prize of $7,500. The flight is noted in block letters on the rear left side fuselage of the aircraft: LOS ANGELES TO CHICAGO-9 HRS. 9 MIN. 4 SEC. AUGUST 27 1930.

Intent on beating the record set by the airship Graf Zeppelin's globe-circling voyage in August 1929 (the circumnavigation, with numerous stops, took twenty-one days, five hours and thirty-one minutes and covered 20,651 miles), Post secured F. C. Hall's backing. Post, however, was not completely well versed in aerial navigation, so to accompany him on the flight he hired the aforementioned Harold Gatty.

For this flight, the *Winnie Mae* was equipped with a roof hatch for celestial observation and an opening on the side of the fuselage for the drift indicator Gatty had developed to determine the aircraft's ground speed and wind correction angles. The aircraft also benefited from a recently developed Sperry artificial horizon and directional gyro.

On June 23, 1931, the pair set out from Roosevelt Field on Long Island, New York, on a 15,000-mile, fourteen-stop journey that would take them across Europe and Russia, including Siberia. On the transatlantic leg of the flight they made a record-breaking crossing in sixteen hours, seventeen minutes. The flight continued to Berlin, Moscow, and Khabarovsk in the Russian Far East. From there, Post and Gatty crossed the Bering Sea, landing in Solomon, Alaska, before forging on to Edmonton, Alberta. They returned to Roosevelt Field after an eight-day, fifteen-hour, fifty-one-minute flight—twelve days faster than the previous record set by *Graf Zeppelin*.

Wiley Post wears the pressure suit he helped to design for high-altitude flight in the Lockheed 5C Vega *Winnie Mae*.

On July 2 they were given a ticker-tape parade in New York City, and on July 6 they were honored by President Herbert Hoover with a luncheon at the White House. That same year Rand McNally & Company published a ghostwritten account of the historic flight (attributed to Post and Gatty) titled *Around the World in Eight Days*, with an introduction by the American humorist Will Rogers, who was a fellow Oklahoman and had become Post's friend after the record-setting flight.

Two years later Post decided to fly the *Winnie Mae* around the world in a solo attempt to break the record that he and Gatty had set. He installed an autopilot device and a radio direction finder that were in the latter stages of development by Sperry and the US Army Air Corps.

On July 15, 1933, he took off from Floyd Bennett Field in Jamaica Bay, Brooklyn, New York, heading for Berlin, where repairs were made to his malfunctioning autopilot. From there he went on to Königsberg, East Prussia, where he obtained maps that he had forgotten in the United States. From Moscow (more autopilot repairs), Post continued to Novosibirsk and Irkutsk, Siberia (still more autopilot repairs), then to Rukhlovo and Khabarovsk, Siberia, before continuing to Flat, Alaska (propeller replaced), Fairbanks, Alaska, and Edmonton, Alberta.

From there Post flew to Cleveland, Ohio, and, on July 22, arrived back at Floyd Bennett Field just seven days, eighteen hours, and forty-nine minutes after he had left, beating his previous record by twenty-one hours. He was greeted by a crowd of nearly fifty thousand people.

On August 1, 1933, New York City honored Post with yet another ticker-tape parade. The next day Post and his wife, Mae, traveled by train to Washington, D.C., where they met with President Franklin D. Roosevelt, who thanked the pilot for his "courage and stamina."

Bottom left: Harold Gatty (left) and Wiley Post with the Lockheed 5C Vega *Winnie Mae* during their record-breaking around-the-world flight in 1931.

Bottom right: *Winnie Mae* is shown in flight.

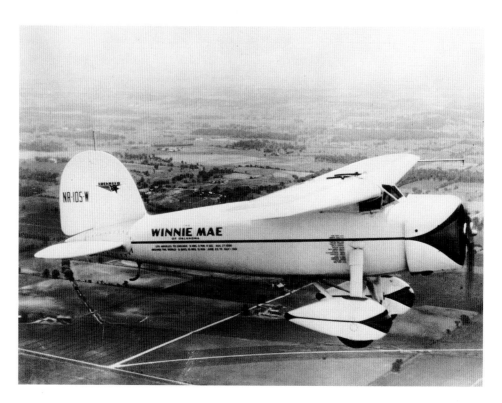

On August 24, 1933, Post was awarded the International Harmon Trophy as the world's outstanding aviator at the Century of Progress International Exposition in Chicago.

Post's two around-the-world flights helped to make him a Depression-era hero and celebrity in America. With these accomplishments under his belt, Post decided to take *Winnie Mae* to new heights. In 1934, he began to probe the possibilities of high-altitude, long-distance flying. The *Winnie Mae*'s cabin was not pressurized, however, driving Post, with the help of the B. F. Goodrich Company, to develop the world's first pressured flight suit. He attempted numerous times in 1935 to set solo high-altitude transcontinental speed records, but none were successful. One particular attempt on March 15, however, was noteworthy. Wearing the pressurized suit and flying at an altitude of more than 30,000 feet, Post flew the *Winnie Mae*—now equipped with a supercharger and jettisonable landing gear—from Burbank, California, to Cleveland, a distance of 2,035 miles in seven hours and nineteen minutes. At times the aircraft reached a ground speed of 340 miles per hour, showing that significant speed increases could be achieved by flying at high altitudes.

In August 1935, Post and Will Rogers set out on an aerial tour of Alaska and Siberia. Post was flying a hybrid aircraft made from a Lockheed Orion and a Lockheed Explorer, overly heavy with fuel plus hunting and fishing gear. The plane was powered by a 550-horsepower Pratt & Whitney Wasp engine and fitted with floats that were reportedly ill-suited for the already makeshift aircraft. On August 15, the pair left Fairbanks, Alaska, bound for Point Barrow. Flying in fog, Post got lost and was forced to land in a lagoon to get his bearings. When they took off again, the engine failed and the aircraft plunged to the ground. Both men were killed instantly.

The deaths of Post and Rogers brought about an international outpouring of grief. Post's body lay in state in the rotunda of the Oklahoma state capitol building, and among his distinguished mourners were famous American aviators such as Amelia Earhart and high-ranking political figures.

Following Post's death, the US Congress purchased the *Winnie Mae* on behalf of the Smithsonian and to relieve the destitute Mae Post of her debts.

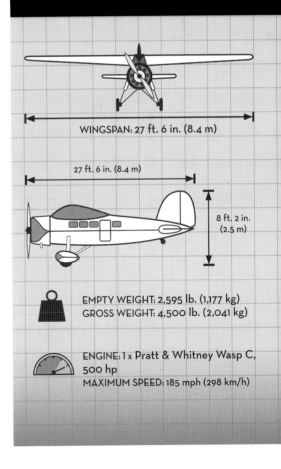

SPECIFICATIONS
LOCKHEED VEGA
5C *WINNIE MAE*

WINGSPAN: 27 ft. 6 in. (8.4 m)

27 ft. 6 in. (8.4 m)

8 ft. 2 in. (2.5 m)

EMPTY WEIGHT: 2,595 lb. (1,177 kg)
GROSS WEIGHT: 4,500 lb. (2,041 kg)

ENGINE: 1 x Pratt & Whitney Wasp C, 500 hp
MAXIMUM SPEED: 185 mph (298 km/h)

PIPER J-3 CUB

CHAPTER 8

BY DOROTHY S. COCHRANE

The Piper Cub brings to mind a little yellow airplane with a smiling cub logo on its tail skimming the treetops and landing in a grassy field on a warm summer day. It embodies the personal dream of flight—a simple and inexpensive machine with gentle flying characteristics. Created as a trainer to foster aviation in the depths of the Depression, the Cub became a ubiquitous utility plane of World War II. Since 1932, thousands of pilots have learned to fly in Cubs, and in the twenty-first century thousands more continue to fly them, along with their light sport and bush plane derivatives.

Two men were responsible for the Cub's success: C. G. Taylor, who designed it, and William Piper, who provided financial stability and marketing genius. By 1941, one-third of all general aviation aircraft were Taylor or Piper Cubs, and more than twenty-seven thousand were built before production ended in 1947. The Cub captured, even arguably created, the private pilot market and remains one of the most recognized designs in aviation. If any aircraft can be anointed the "generic light plane," the Piper Cub is it.

In the 1920s, the public became increasingly air minded as it reveled in barnstorming, air racing, and the adventures of globetrotting aerial explorers. Relentlessly touted by true believers, aviation took on a gospel tone that evolved into a more practical and productive time known as the

The Smithsonian National Air and Space Museum's Piper Cub J-3, one of the most recognizable aircraft in the world, hangs in the Steven F. Udvar-Hazy Center.

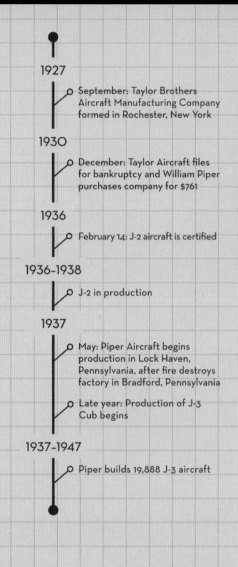

1927

○ September: Taylor Brothers Aircraft Manufacturing Company formed in Rochester, New York

1930

○ December: Taylor Aircraft files for bankruptcy and William Piper purchases company for $761

1936

○ February 14: J-2 aircraft is certified

1936–1938

○ J-2 in production

1937

○ May: Piper Aircraft begins production in Lock Haven, Pennsylvania, after fire destroys factory in Bradford, Pennsylvania

○ Late year: Production of J-3 Cub begins

1937–1947

○ Piper builds 19,888 J-3 aircraft

Golden Age of Aviation. The federal government built a regulatory framework while Charles Lindbergh and Amelia Earhart ignited aviation fever with their respective record-setting flights of 1927 and 1928. Although private flying was, in reality, mainly for working pilots or wealthy sportsmen who flew large and expensive planes, homebuilt and light planes, such as the Heath Parasol and Pietenpol Air Camper, and powered glider-style craft, such as the Aeronca C-2, soon appeared. Ultimately, it was the Taylor/Piper Cubs that offered flight schools, clubs, and the general public what they wanted: a truly practical, affordable, and fun light airplane.

In 1911, at the age of thirteen, Clarence Gilbert Taylor marveled at the Wright EX *Vin Fiz* flying over upstate New York on the first transcontinental flight across the United States. As a teenager, Taylor built his first design in the attic of his home in Rochester, New York, and became a self-taught engineer at the Tool, Die and Specialty Company with his father, Arthur, and brother, Gordon. In 1926 C. G. and Gordon formed a Curtiss Jenny barnstorming venture, then founded the Taylor Brothers Aircraft Company and built the Arrowing A-2 Chummy, a small parasol-wing, side-by-side, open-cockpit monoplane with a radial engine. Tragedy struck in April 1928, when Gordon was killed in the A-2 at an exhibition in Detroit, Michigan, but C. G. continued his work. Looking for larger quarters, he moved to Bradford, Pennsylvania, at the invitation of the chamber of commerce, which also lured oilman William Piper and a partner to pay $400 to become original investors in the Taylor Company. The Bradford investors exemplified the 1929 aviation boom, a pivotal moment of municipal and Wall Street investment in aircraft development.

Recognizing that truly simple and inexpensive aircraft were scarce, entrepreneur Harry Guggenheim organized the International Safe Aircraft Competition, sponsored by the Daniel Guggenheim Fund for the Promotion of Aeronautics, for aspiring small-plane designers. Taylor entered the Chummy, which, along with six other aircraft, failed to qualify; five other planes withdrew. However, the winning plane, the Curtiss Tanager, proved too expensive to build. The intense moment of opportunity proved short lived when the Wall Street bubble burst in October 1929 and wiped out many designs within a year. The expensive Chummy, with a $4,000 price tag, and Taylor Airplane Company were no exceptions.

In 1930, investor William Piper, now intrigued with aviation, bought the Taylor Company for $761, installed himself as treasurer, and retained Taylor as president and chief engineer. Piper understood that contemporary private aircraft were affordable only to the rich, were really working airplanes, or were so underpowered they barely flew at all. He and Taylor were determined to build an aircraft that was inexpensive to rent or own and stable enough to learn to fly. He was not alone—even in the worsening economy, others hoped to accomplish the same.

The Aeronca C-2 was the first truly light airplane certified by the Department of Commerce and the Bureau of Aeronautics and produced in substantial numbers in the United States. Safe, economical, and easy to fly, this open-cockpit monoplane sold for under $1,300 and could be rented for only $4 an hour. (The National Air and Space Museum's two-seat Aeronca C-3, the production prototype, first flew in October 1929.) The moderately priced Curtiss-Wright CW-1 Junior—with short-field capability, low handling speed, and good visibility with the engine behind the pilot—became a one-year wonder before production ceased (but also earned a place in the NASM collection). Others in the market included the American Eagle Eaglet 31 and Buhl Bull LA-1 Pup.

Taylor first built the high-wing D-1 glider with a steel tube frame and a tailskid. He and Piper pulled it behind a car to flight test it, while Piper's son Tony, noting its size, commented, "You just put it on." Next, broadly drawing from the D-1, Taylor designed a high-wing, tandem-fuselage powered glider in which was installed a 20-horsepower Brownback Tiger Kitten engine. The engine was so weak that the plane indulged only in grass-cutting along the ground, but it motivated Taylor's accountant, Ted Weld, to observe that an airplane with a Kitten engine should be known as a Cub. Piper initially installed an expensive French Salmson radial engine, but the cost and weight were simply impractical for a small plane. His search for an alternative led him to the Continental Aircraft Engine Company, which was creating a radically new engine with four cylinders laid out in two horizontally opposed banks. Besides lower weight and cost, the "flat" 40-horsepower engine offered better visibility and less drag. Piper became an early customer, and the engine was crucial to the ultimate success of the Cub and the civilian light-aircraft industry. The Cub had arrived, and a legend was born.

Taylor's E-2 Cub received certification in 1931 with a 37-horsepower Continental A-40-2 engine. The fabric-covered wing of Sitka spruce spar and aluminum ribs had a popular USA-35B airfoil and connected to the tubular steel fabric-covered framework to form a semi-enclosed cabin. Steel-tube struts and rubber shock-cord landing-gear springs made the Cub sturdy, and a hinged side-panel door allowed for relatively easy entry. The only instruments were an altimeter, a tachometer, and oil temperature and pressure gauges; a wire-and-cork fuel gauge bobbed up from the tanks in the wing roots. Other trial engines resulted in F, G, and H models, but the E-2 struck a chord. It weighed only 925 pounds and sold for $1,325, compared to $1,730 for the Aeronca C-3 and $1,595 for the Curtiss-Wright

Opposite: This air-to-air view of the classic Piper Cub was created by renowned aviation photographer Hans Groenhoff.

Below: The spartan interior of a Piper J-3 Cub features only basic instruments: altimeter, air-speed indicator, oil-pressure and temperature gauges, and tachometer. This is true "seat-of-the-pants" flying.

HOW TO FLY

A PRACTICAL FLIGHT-MANUAL FOR MEMBERS OF

CUB PILOT CORPS

The Cub Pilots Corps was a marketing tool created by William Piper to encourage the public to learn to fly and enjoy the camaraderie and low costs of flying clubs.

Junior. Piper's financing kept the company in business until E-2 sales finally picked up in 1934; more than three hundred were sold between 1931 and 1936.

Still not satisfied with the design, and acting on owner suggestions, Piper hired engineer Walter Jamouneau in 1935 to improve the aircraft. Jamouneau designed a true cabin by raising the turtle deck and fairing it into the trailing edge of the wing, rounding the wingtips, and giving the little plane a curved rudder and fin. Skipping the letter I, the Taylor J-2 emerged with the 40-horsepower Continental A-40-4 engine and a Sensenich wooden propeller that offered a cruise speed of only 70 miles per hour but nice glide qualities and gentle stalls. Wheel landings taught proper use of the rudder for the pilot, who sat in the rear seat for solo flight. With few hard-surface airports yet available, the Cub could be fitted with wheels, floats, or skis. The fixed vertical stabilizer on early models sported the brand name "The Cub." The J-2 cost $1,470 and rented for $10 an hour.

William Piper approved the changes; however, Taylor's objections to the new design led to the final breakdown of the partnership and Piper bought out Taylor in December 1935. Taking his father and a few employees with him, Taylor moved to Butler, Pennsylvania, and then to Alliance, Ohio, where he built a light plane with the improvements he had sought for the Cub, including side-by-side seats and a new airfoil. The Taylorcraft Model A received certification in June 1937 and became a worthy competitor of the Cub.

About 695 Taylor J-2s were produced at Bradford until the factory burned to the ground in March 1937. Piper moved the company to an old silk factory in Lock Haven, Pennsylvania, and restarted production in July with jigs and tools saved from the fire. Piper installed modern assembly lines in the large facility and, in November, changed the company's name to Piper Aircraft Corporation. Overall the factory built 1,547 E-2 and J-2 Cubs from February 1931 to May 1938.

To compete with Taylorcraft and Aeronca, Piper introduced the J-3 in late 1937 with "luxuries" now considered necessities: brakes, a steerable tail wheel, upholstered (instead of plywood) seats, a compass, an airspeed indicator, and more leg room due to the firewall being moved forward. Priced at $1,300, then reduced to $995 in 1939, it was a resounding success with fixed-base operators, flight schools, and air-minded citizens who, due to easing economic times, could now afford to learn to fly. Piper built about 19,888 J-3s from 1938 to 1947.

William Piper showed his marketing savvy in several ways, foremost in adopting the automobile purchasing model—he charged $425 down and twelve monthly payments for the J-3, which allowed a plane to be simultaneously flown and paid for at flying schools. He established dealerships and clubs, distributed "How to Fly" kits and pins, and sponsored model-building contests. Piper branded his little plane by standardizing the color to Lock Haven Yellow—an orange-yellow soon replaced by a lighter Piper Cub chrome yellow with black trim and the now-familiar cub on the tail. Salesmen barnstormed by day and moved to a new city each night. Marketing to universities with National

Intercollegiate Flying Clubs allowed the company to target young college students (and future buyers) who competed in air meets. Cubs were perfect for stunts at air races and air shows; slow and maneuverable, they could take off from and land on a platform attached to a moving car or allow a "flying farmer" (a disguised pilot) to "steal" one and awkwardly perform for a skeptical crowd.

As the Depression eased, private flying increased. More than seven thousand new private aircraft were produced in 1939. The Civilian Pilot Training Program (CPTP), established in July 1939 to provide basic flight training to men in preparation for possible military service, spurred sales of the J-3 to 3,016 in 1940. Seventy-five percent of all pilots in the CPTP trained in Cubs.

In 1940, while ramping up for wartime aircraft production, the military foresaw no need for light planes, forcing Piper and his competitors to request field maneuvers to demonstrate the category's ability to move personnel and perform observations duties for less cost and easier field maintenance. The Cub's tandem design, similar to that of military training aircraft, appealed to the army air forces, which accepted it in 1942.

During World War II, more than 5,600 Cubs flew in and out of short fields as liaison, observation, and ambulance airplanes in support of invasions of North Africa, Europe, and in the Pacific, hedgehopping over battlefields or ferrying officers, including Supreme Allied Commander Gen. Dwight Eisenhower. Nicknamed "Grasshopper" because it leaped from the

The Piper Cub's short-field capabilities allowed the US military to utilize thousands of Piper L-4s, nicknamed "Grasshoppers" because they leapt off the field to provide transport, observation, and liaison duties during World War II.

Above: The "low and slow" Cub is perfect for crop-dusting farm fields.

Opposite: The Piper Cub legacy continues today in light sport and backcountry designs like this Cub Crafters CC18-180 Top Cub that are simple, rugged, and fun to fly.

ground like an insect, the military Cub carried several official designations: L-4, O-59, navy HE-1 and AE-1, and marine NE-1 and NE-2. (The NASM's military Cub is an L-4B with serial number 43-1074.)

Piper Cubs also flew with foreign air forces and in World War II and the Korean War. Dramatic Grasshopper stories include that of the Piper L-4 *Miss Me*, credited with the last aerial victory in Europe in late April 1945, achieved when its crew used pistols to bring down a German Fieseler Storch. At wartime peak, a new Cub emerged from the factory every twenty minutes, many ferried by the Women Airforce Service Pilots (WASP) from Lock Haven to military airfields; some civilian pilots flew their Cubs in the Civil Air Patrol (CAP).

Piper ended J-3 production in 1947 in favor of models with improved horsepower and amenities, beginning with the side-by-side J-4 Coupe and the three-seat J-5 Cruiser. Two Piper PA-12 Super Cruisers, a more powerful version of the J-5, became the first light planes flown around the world: from August 9 to December 10, 1947, Clifford Evans in the *City of Washington* and George Truman in the *City of The Angels* flew 22,436 miles. (The planes are now found at the Steven F. Udvar-Hazy Center and the Piper Aviation Museum in Lock Haven, Pennsylvania, respectively.) In 1966, teenage brothers Rinker and Kernahan Buck rebuilt a PA-11 Cub Special and became the youngest aviators to make a US transcontinental flight, immortalized in Rinker's classic memoir *Flight of Passage*.

While the J-3 is the most storied version of the Cub, it may be that the descendant Super Cub is today's most popular and enduring one. Introduced in 1949, the PA-18 was a strengthened PA-11 with a 150-horsepower engine but without the famous yellow color or cub logo. About 8,500 civilian and 1,500 military Super Cubs were built at Lock Haven and in Vero Beach, Florida. In May 1951, aerobatic champion Betty Skelton flew one to a world light-plane altitude record of 29,050 feet. Super Cubs are revered in the backcountry because they combine the classic Cub low stall speed and short-field capability with rugged construction and adaptability for bush tires or floats, bigger engines and propellers, and new avionics. The Atomic Energy Commission (AEC) Super Cub now in the NASM collection was used for uranium exploration in the remote Southwest.

William Piper and the board of directors anointed serial number 1937 as the first official Piper J-2 and flew it as the company plane for two years. Hal Goff of Aero Enterprises eventually found the airplane in a hangar in West Virginia and restored it to airworthy condition. He flew it to Lock Haven for the 1976 rollout ceremony of the one hundred thousandth Piper aircraft (a Cheyenne), where Jamoneau and astronaut Pete Conrad flew the J-2. Its last owner, Lefferts Mabie Jr., later donated it to the NASM in December 1984.

The National Air and Space Museum's 1941 Piper J-3 Cub, serial number 6578 (originally designated a J3L-65), spent time in numerous places around the South, including Jonesboro, Arkansas, where it likely was a trainer in the CPTP. Roland M. Howard of Houston, Texas, restored it and with friends donated it to the NASM with 5,655 flight hours in 1977.

The Cub deeply influenced the rise of private flying and the light plane industry in the United States. Its easy transition from civilian to military, utility, and bush plane sealed its place in history as the generic light plane that remains relevant today. The simple romance of flight that defined its birth still appeals, and the beloved Cub design is still pure fun to fly. C. G. Taylor and William Piper would be eminently pleased, and possibly not surprised, to find their Depression-era visions of a practical light aircraft still thriving in the twenty-first century.

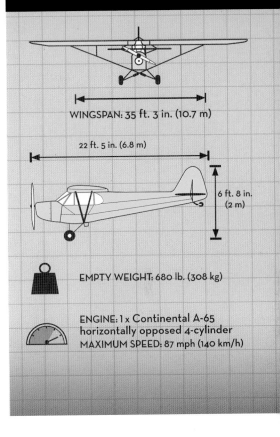

SPECIFICATIONS
PIPER J-3 CUB

WINGSPAN: 35 ft. 3 in. (10.7 m)

22 ft. 5 in. (6.8 m)

6 ft. 8 in. (2 m)

EMPTY WEIGHT: 680 lb. (308 kg)

ENGINE: 1 x Continental A-65 horizontally opposed 4-cylinder
MAXIMUM SPEED: 87 mph (140 km/h)

CHAPTER 9

BY TOM D. CROUCH

At 11:40 Mountain Standard Time on the morning of November 11, 1935, US Army Air Corps Capts. Albert W. Stevens and Orvil A. Anderson were higher than any other human beings had ever been—or would be again for another twenty-one years. At an altitude of 72,395 feet, the atmospheric pressure was so low that their balloon, which at launch had been tall and thin with a bubble of helium at the very top, was now completely round. Peering out the portholes of their pressurized gondola, they could see the curvature of Earth. The land spread out beneath them, Stevens noted, was "a vast expanse of brown, apparently flat, stretching on and on." The sky shaded from light blue at the horizon to "black, with the merest suspicion of a very dark blue," overhead.

Until the Space Age, the balloon, the oldest means of taking to the air, continued to offer one great advantage over the airplane: the ability to climb to the very roof of the atmosphere, where there was too little air to provide lift for wings or to support air-breathing engines. As a result, balloons remain today essential tools that enable scientists to study conditions in the upper atmosphere and near-Earth space and to develop and test equipment and techniques to protect future generations of space travelers.

Today the *Explorer II* gondola can be viewed in the Smithsonian National Air and Space Museum's Barron Hilton Pioneers of Flight Gallery.

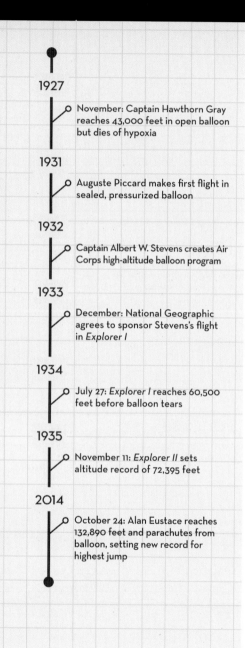

1927

November: Captain Hawthorn Gray reaches 43,000 feet in open balloon but dies of hypoxia

1931

Auguste Piccard makes first flight in sealed, pressurized balloon

1932

Captain Albert W. Stevens creates Air Corps high-altitude balloon program

1933

December: National Geographic agrees to sponsor Stevens's flight in *Explorer I*

1934

July 27: *Explorer I* reaches 60,500 feet before balloon tears

1935

November 11: *Explorer II* sets altitude record of 72,395 feet

2014

October 24: Alan Eustace reaches 132,890 feet and parachutes from balloon, setting new record for highest jump

John Jeffries, a loyalist American physician who served the British during the Revolutionary War, launched the era of balloon science with two flights accompanied by French aeronaut Jean-Pierre Blanchard—including, on January 7, 1785, the first aerial voyage across the English Channel. In preparation for his adventure, Dr. Jeffries commissioned a London instrument maker to provide a large barometer and thermometer, which he used to record conditions aloft. The world's first aeronautical instruments, they are displayed at the National Air and Space Museum.

Throughout the nineteenth century and into the twentieth, balloonists continued to brave frigid temperatures, low oxygen levels, and falling atmospheric pressure in order to push higher into the sky. Some, like Capt. Hawthorne C. Gray of the US Army Air Corps, paid for their daring with their lives. Ascending to an altitude over 43,000 feet in November 1927 to test clothing and equipment that would enable aviators to survive and fight a future war in the upper air, Gray accidently died of hypoxia. His open balloon basket is displayed in the National Air and Space Museum's Pioneers of Flight gallery.

The desire to study the cosmic radiation showering the Earth inspired a new era in high-altitude ballooning. Swiss physicist Auguste Piccard made two flights in the first sealed, pressurized balloon gondola in 1931 and 1932. Ascending to altitudes above 50,000 feet, he conducted his research in relative safety above most of the filtering atmosphere, returning from the stratosphere as a scientific celebrity of the first order.

Captain Albert W. Stevens recognized that ballooning to a record altitude would accomplish important technological and scientific goals while achieving international prestige for the air corps and the nation. Twice decorated for valor while serving as an aerial photographer and bombardier during World War I, Stevens returned to the United States as a captain assigned to the air corps engineering division at McCook and Wright Fields in Dayton, Ohio. He spent the next two decades developing the equipment and techniques required to take high-quality oblique aerial photographs. He also participated in several large-scale aerial photographic expeditions and was involved in a record-setting altitude flight during which he parachuted from 24,306 feet.

In 1932, Stevens launched an effort to establish an air corps high-altitude balloon program. Piccard's balloon flights had inspired Soviet dictator Joseph Stalin to send his airmen to record heights as a means of building national pride and impressing the rest of the world with USSR technological prowess. The US Navy entered the field as well, beating the Soviets with a balloon flight to over 61,000 feet. The competition to set an ever higher mark came at a cost: on January 10, 1934, a Soviet crew rose to a new record altitude only to lose their lives when their gondola broke loose and smashed to Earth.

While air corps leaders gave permission for Stevens to begin planning a flight, they declined to fund the development or manufacture of the balloon, gondola, or other essential pieces of equipment. Stevens turned to his friend Gilbert Grosvenor, head of the National Geographic Society. Always looking for a potential magazine story combining adventure with the trappings of science and a whiff of danger, Grosvenor saw that Stevens and his project were made to order for the National Geographic Society, which in December 1933, the society agreed to donate $25,000 to support the enterprise. Donations from other individuals and corporations brought the total to $45,000. Dow Chemical agreed to manufacture the gondola using a lightweight magnesium alloy, and Goodyear-Zeppelin the balloon envelope, at well below cost.

Captains Stevens and Anderson and Maj. William E. Kepner, the crew of *Explorer I*, rose out of the Stratobowl, their South Dakota launch site, at 7:50 a.m. on July 27, 1934.

By noon they had reached an altitude of 40,680 feet. An hour later they were approaching 60,500 feet when they noticed that several large rips were developing in the lower portion of the envelope. Beginning their descent without having reached a record altitude, they were passing through 16,000 feet when the hydrogen-filled gas bag began to disintegrate.

With their balloon plummeting toward Earth, Kepner opened the hatch and climbed onto the top of the gondola to assist the others in exiting. Anderson's parachute opened when he was only halfway out of the hatch, and he jumped holding the silk canopy in his arms. The pressure of the wind trapped Stevens halfway out of the hatch for a few seconds, until he was finally able to push free. Minutes later, all three men were safely on the ground. The National Geographic Society, which had gotten a much more exciting story than it would have if the flight had been a complete success, agreed to fully fund another attempt.

The resulting *Explorer II* balloon boasted a capacity of 3,700,000 cubic feet—700,000 cubic feet more than its predecessor. Instead of hydrogen, it would be filled with slightly heavier but much safer helium gas. The new gondola featured hatches that were 4 inches larger in diameter than those on *Explorer I*, a result of lessons learned. With Kepner now assigned to other duties, Stevens and Anderson would make this flight without him. The first attempt to inflate the *Explorer II* balloon on the floor of the Stratobowl on July 12, 1935, failed when a faulty ripping panel, designed to allow rapid deflation at the end of the flight, opened prematurely. Repairs and the installation of an improved design delayed another attempt until November 11. Four and a half hours after takeoff they were looking out on the world from 72,395 feet. The rest of the flight was uneventful, with a safe landing near White Lake, South Dakota, at 3:14 p.m.

The *Explorer* program was far more than a stunt. "Among the sixty-three separate studies conducted," Anderson later reported, "cosmic radiation both in density and direction was one of the more important." Radiation at altitude was eighty times that recorded at takeoff.

Bottom left: *Explorer II* is ready for launch from the Stratobowl in South Dakota on November 11, 1935. It would ascend to 72,395 feet.

Bottom right: Interior view of the *Explorer II* gondola.

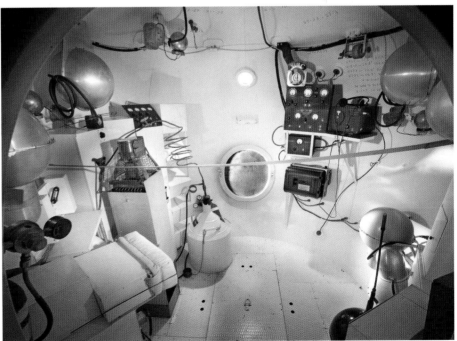

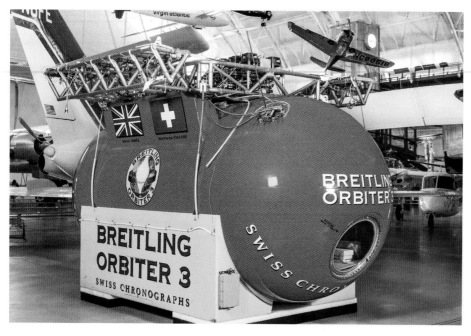

Top left: The gondola of *Breitling Orbiter 3*. In March 1999, Bertrand Piccard and Brian Jones piloted the *Orbiter* on the first nonstop circumnavigation of the globe in a balloon.

Top right: Steve Fossett completed the first nonstop solo flight around the world in a balloon aboard *Bud Light Spirit of Freedom* on June 19 to July 24, 2002.

The pair exposed spores and fruit-fly larvae to cosmic rays to study their impact on living organisms and brought samples of the atmosphere back for analysis.

On the twentieth anniversary of the historic flight, Dr. Hugh L. Dryden, director of the National Advisory Committee for Aeronautics (NACA), noted that in addition to their very real scientific contributions, the flights were a "convincing demonstration that man could protect himself from the environment of the stratosphere." General Henry "Hap" Arnold, commander of the US Army Air Forces during World War II, remarked that the Stratobowl flights made significant contributions to the Allied victory, including the development of improved magnesium alloys, pressurization gear, and flight clothing. "Many other items of equipment and methods were improved," General Arnold continued, "which later played important parts in giving American airmen superiority in the skies of Berlin and Tokyo." Visitors to the National Air and Space Museum today can see the historic *Explorer II* gondola in the Pioneers of Flight Gallery.

After World War II, with aircraft flying ever higher and the possibility of space flight on the horizon, balloons once again paved the way to extreme altitudes. The US Navy Strato-Lab and US Air Force Manhigh balloon projects took turns establishing new records and testing the abilities of human beings and their equipment to function on the edge of space. Could a pilot survive a parachute jump from the stratosphere? US Air Force Capt. Joseph Kittinger proved that he could, taking the *Excelsior III* balloon to an altitude of 102,800 feet over the New Mexican desert and parachuting safely to earth on August 16, 1960.

Nick Piantanida, a New Jersey truck driver and skydiver, was determined to challenge Kittinger's record. On February 2, 1966, he ballooned to 123,500 feet in an unpressurized gondola, but a frozen oxygen valve prevented him from jumping. He lost his life on the next attempt, on May 1, 1966, probably as a result of an attempt to clear the faceplate of his pressure helmet. The gondola of his *Strato-Jump III* is displayed at the NASM's Steven F. Udvar-Hazy Center.

Austrian Felix Baumgartner, a veteran skydiver, finally broke Kittinger's mark on October 14, 2012, when he rose to 128,100 feet over Roswell, New Mexico, in his pressurized

Red Bull Stratos gondola and jumped. In the process, he set a new balloon altitude record and records for the highest altitude parachute jump and the longest time in free fall. And while still in the high, thin air near the top of his jump, with the whole world watching on live television, Baumgartner became the first human being to break the speed of sound in free fall. The project's success was the result of the efforts of a strong project team. Art Thompson, founder of Sage Cheshire Aerospace Inc. headed the engineers and technicians who designed, built, and tested the high-tech capsule; served as flight test director; and selected the other members of the *Red Bull Stratos* team. Colonel Joe Kittinger, the previous record holder, mentored Baumgartner and handled all communications with the capsule, while Dr. John Clark, who served as crew surgeon on six Space Shuttle missions, was project medical director. "The aim," Dr. Clark reported, "is to improve the safety for space professionals as well as potential space tourists." The *Red Bull Stratos* gondola joined the other balloon treasures on display at the Udvar-Center.

Felix Baumgartner's records would not stand as long as those he had shattered, however. On October 24, 2014, Senior Vice-President of Knowledge Alan Eustace of Google was carried aloft by a plastic balloon that stood almost 450 feet tall at launch and expanded to a diameter of 311 feet at altitude. Two and one-half hours after launch, Eustace reached an altitude of 135,890 feet. He wore an advanced pressure suit designed and produced by ILC Dover Industries, coupled with a life-support system developed by Paragon. United Parachute Technologies provided the drogue, main, and reserve parachutes. ADE Aerospace Consulting supplied the medical team and required training, with Dr. Clark serving as medical consultant. Julian Nott, the project's ballooning consultant, proposed a unique launch procedure. The Tata Institute of Fundamental Research, Hyderabad, India, produced all of the balloons used in the program.

From the outset, the Project StratEx team saw little point in spending a lot of money to develop a sophisticated and heavy pressurized cabin. Instead, Eustace made the flight dangling beneath the balloon via a beam structure known as the Balloon Equipment Module, protected only by his pressure suit, for a total weight of just 450 pounds. Not long after dropping from the balloon, Eustace reached a record top speed of 822 miles per hour—Mach 1.23. In addition, he set new world records for the highest jump (135,890 feet) and for distance of fall using a drogue chute (123,414 feet). Just fifteen minutes after cutting loose from the balloon, Eustace was safely back on the ground. In the best NASM tradition, a curatorial team representing both the aeronautics and space history divisions was able to acquire the pressure suit, parachute, and other gear for the NASM collection.

In addition to displaying the gondolas and baskets of balloons that have climbed to the roof of the sky, the National Air and Space Museum preserves and exhibits lighter-than-air craft that have flown across oceans and around the globe. On August 11, 1978, Maxie Anderson, Ben Abruzzo, and Larry Newman lifted off from Presque Isle, Maine, aboard the *Double Eagle II*. One hundred and thirty-seven hours and six minutes later the trio made a safe landing in Miserey, France, having completed the first balloon crossing of the Atlantic. From March 1 to 21, 1999, Bertrand Piccard and Brian Jones piloted *Breitling Orbiter 3* on the first nonstop circumnavigation of the globe, while Steve Fossett completed the first nonstop solo flight around the world aboard *Bud Light Spirit of Freedom*, on June 19 to July 24, 2002. All are represented in the NASM collection.

No other museum in the world can offer visitors such a collection of balloons that have climbed so high or flown so far.

SPECIFICATIONS
EXPLORER II

GONDOLA DIAMETER: 9 ft. (2.8 m)

GROSS WEIGHT: 640 lb. (290 kg)
EMPTY WEIGHT: 640 lb. (290 kg)
PAYLOAD: 1,500 lb. (700 kg)

DOUGLAS DC-3

CHAPTER 10

BY DOMINICK A. PISANO

It has become a cliché among aviation historians to call an aircraft an icon, but in the case of the Douglas DC-3, the term is remarkably apt. The first profitable passenger-carrying airliner, it became a functional and striking symbol of its time, a beautiful, streamlined machine that inspired industrial designers to copy its lines in automobiles, locomotives, and consumer goods. Moreover, the DC-3, modified and manufactured in great numbers for wartime use as the C-47 (in multiple variants), went on to widespread service during World War II, earning a reputation as a sturdy workhorse. Long after the war and into the present, modified and converted examples continued to perform a variety of aerial duties.

The DC-3 in the National Air and Space Museum's collection (#344) was owned and operated by Eastern Air Lines from December 1937 to October 1952. The aircraft flew a total of 56,782 hours, the most of the entire Eastern DC-3 fleet. It made its last scheduled flight on October 12, 1952, in honor of Columbus Day, from Miami to El Salvador, in the Bahamas, with veteran Eastern pilot Dick Merrill at the controls. When it returned to Miami, the aircraft was retired from service. In 1953, Eastern donated the aircraft to the National Air Museum. It was placed in storage until late 1974, when it was restored with the help of volunteers from Eastern. The DC-3 was placed on display at the current National Air and Space Museum building, which opened to the public in July 1976.

The National Air and Space Museum's Douglas DC-3, flown by Eastern Air Lines, is on display in the America by Air Gallery.

TIMELINE
DOUGLAS DC-3

1932

○ August 2: Jack Frye letter to manufacturers asks for modern twelve-passenger airliner

1933

○ July 1: First flight of Douglas DC-1

1934

○ May 11: First flight of fourteen-passenger DC-2

○ May 19: TWA begins passenger service with DC-2

○ October 26: Unmodified KLM DC-2 places second in England–Australia Air Race

○ Autumn: American Airlines president C. R. Smith requests sleeper version of DC-2

1935

○ December 17: First flight of Douglas Sleeper-Transport (DST)/DC-3

1936

○ June 25: Douglas DST enters service with American Airlines

1941

○ November: C-47 military transport version first flies

1941–1945

○ C-47 serves as primary Allied transport plane during World War II

1948

○ June: Berlin Airlift begins with C-47s

1952

○ October 12: Last flight of Eastern Air Lines DC-3 #344

1953

○ Eastern donates #344 to Smithsonian

DC-3 #344 is representative of one of the aircraft's iterations—its era as the mainstay modern airliner from its inception in 1935 until its replacement by larger and more powerful aircraft like the Douglas DC-4 and DC-6 and the Lockheed Constellation and Super Constellation in the immediate postwar period, and eventually by the Boeing 707 and Douglas DC-8 jetliners in the late 1950s.

The genesis of the DC-3 began in March 1931 after a Transcontinental and Western Air (TWA—subsequently, Trans World Airlines) Fokker F.10A on a westward flight from Chicago to Los Angeles crashed near Bazaar, Kansas, killing the famed Notre Dame football coach Knute Rockne and seven others onboard. The Aeronautics Branch of the Department of Commerce, a predecessor of the present-day Federal Aviation Administration, determined that prolonged wing flutter (unstable oscillation that often leads to structural damage) had caused the aircraft's interior wooden wing spars and ribs to weaken, in turn causing the F.10A's left wing to break from the fuselage. The Aeronautics Branch grounded all other F.10s and demanded that their operators inspect their aircraft regularly for interior wing damage.

Jack Frye, Transcontinental and Western Air's vice president in charge of operations, realized that the cost and time required for these inspections might put TWA out of business and that the F.10's days were probably numbered. He therefore attempted to obtain a more modern airliner from the Boeing Company, which was in the process of developing the 247, a modern streamlined aircraft built of aluminum alloy that had numerous advanced design features, including supercharged twin engines and retractable landing gear. It was also 50 percent faster than any transport aircraft of its time. (The National Air and Space Museum's Boeing 247-D is on display in the America by Air exhibition.) However, the first run of 247s was intended solely for competitor United Air Lines, a firm controlled by parent company United Aircraft and Transport.

On August 2, 1932, Frye wrote a now-historic letter to various aircraft manufacturers in which he requested "ten or more trimotored transport planes." Frye further requested that the aircraft be of all-metal construction, preferably monoplanes but "combination structure or biplane would be considered." The aircraft should be powered by engines of 500 to 650 horsepower, be able to accommodate "at least 12 passengers with comfortable seats and ample room," and have a minimum top speed of 185 miles per hour at sea level. Finally, Frye specified that "this plane fully loaded must make satisfactory take-offs under good control at any TWA airport on any combination of two engines." Because one of TWA's transcontinental route airports was located in Albuquerque, New Mexico, at an elevation

of nearly 5,000 feet, and with temperatures often 90 degrees Fahrenheit and above, this requirement was considered extremely restrictive, if not impossible, to meet.

Donald Douglas, head of the Douglas Aircraft Company, one recipient of Frye's letter, put two of his top engineers, James H. "Dutch" Kindelberger and Arthur Raymond, to work on a proposal. Kindelberger and Raymond believed that TWA's requirements could be met by a twin-engine transport, and they set out to design an aircraft that would be equal to or better than the Boeing 247.

Douglas dispatched Raymond and general manager Harry Wetzel to New York to discuss the proposal with TWA. Negotiations took more than three weeks. TWA's technical advisor, Charles A. Lindbergh, insisted that the proposed aircraft must be able to take off, climb, and maintain level flight on one engine with a full payload from any airport on the TWA route. Raymond commented that he was 90 percent certain such an aircraft could be built, but that the remaining 10 percent was keeping him awake at night. Kindelberger decided that the only way to find out whether Douglas Aircraft could meet the TWA specification would be to build the aircraft.

The result was the Douglas DC-1, for Douglas Commercial One, registration number X223Y. Like the Boeing 247, it featured a cantilevered wing, which had an internal main spar that carried an aircraft's flight load without need for external struts or bracing. The monocoque fuselage consisted of aluminum alloy sheets riveted to an internal framework of bulkheads (upright partitions) and stringers (longitudinal members), which enabled the aircraft's metal skin to bear aircraft flight loads.

Extensive engineering work began at Douglas's Santa Monica, California, plant. The aircraft's design made use of wind-tunnel tests and mockups of the fuselage and fuel and hydraulics systems. The all-metal DC-1 used an aluminum alloy developed by the Aluminum Company of America (Alcoa) called 24SRT Alclad. To meet the takeoff and level-flight requirements, variable-pitch propellers were installed that enabled the pilot to adjust the propellers' angle of attack for various high- and low-speed flight regimes. There was a competition between Wright and Pratt & Whitney for which engines would power the DC-1, but eventually the Wright SGR-1820-F Cyclone nine-cylinder air-cooled radial engine was selected along with three-bladed Hamilton Standard propellers. In addition, the aircraft's cantilevered wing was integrally built under the fuselage so that, unlike the Boeing 247, passengers did not have to negotiate a raised area on the fuselage floor where the wing was positioned.

The twelve-passenger DC-1 completed its maiden flight on July 1, 1933, from Clover Field in Santa Monica, California, with TWA test pilot Carl A. Cover at the controls. Afterward, the DC-1 (now carrying the civil aviation designation NC223Y) was extensively tested for about

Top left: A cockpit view of the Douglas DC-3 shows the instrument panel and pilot and co-pilot controls.

Top right: Passengers enjoy an in-flight meal in the cabin of a TWA Douglas DC-2.

Opposite: An Eastern Air Lines Douglas DC-3 in flight.

DOUGLAS DC-3

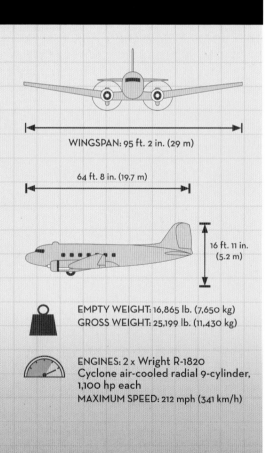

WINGSPAN: 95 ft. 2 in. (29 m)

64 ft. 8 in. (19.7 m)

16 ft. 11 in. (5.2 m)

EMPTY WEIGHT: 16,865 lb. (7,650 kg)
GROSS WEIGHT: 25,199 lb. (11,430 kg)

ENGINES: 2 x Wright R-1820
Cyclone air-cooled radial 9-cylinder,
1,100 hp each
MAXIMUM SPEED: 212 mph (341 km/h)

three months by pilots from Douglas Aircraft, TWA, and the Bureau of Air Commerce before its delivery to TWA in December 1933. TWA used it as a flying laboratory over all parts of its route network, and sometimes it was operated with a few scheduled passengers.

Only one DC-1 was built, and it was eventually sold to Howard Hughes, who wished to use it for various long-distance flight attempts, including an around-the-world flight. Hughes eventually gave up the idea and instead flew a twin-engine Lockheed 14 Super Electra in his successful circumnavigation of the globe in July 1938.

The DC-1 exchanged hands again, until it became part of the Iberia fleet. In December 1940, it crashed on takeoff during a scheduled flight from Seville to Málaga, Spain, with a final destination of Tétouan, Morocco. Although no one was killed, the aircraft was damaged beyond repair.

While the DC-1 was in development and test-flight stages, it was discovered that the aircraft was unstable in yaw. To correct that problem, Douglas lengthened the aircraft, giving it the added benefit of an extra row of seats. By this time TWA realized that a larger aircraft was required and put in an order for twenty fourteen-passenger DC-2s. In addition to the commercial version, Douglas made military transport types for the US Army Air Corps (the C-33) and for the US Navy (the R2D-1). A number of DC-2s were converted to military use during the early days of World War II.

Initial deliveries of the commercial DC-2 went to TWA, American Airlines, Pan American Airways, Pan-American Grace Airways, Eastern Air Lines, and General Air Lines (Western Air Express) in the United States. Among the European airlines that purchased the DC-2 was KLM (or Royal Dutch Airlines), based in Amsterdam. On May 19, 1934, shortly after the DC-2's maiden flight on May 11, TWA took possession of its first DC-2 (*City of Chicago*) and used it on its Columbus–Pittsburgh–Newark route.

The DC-2 made its mark in the MacRobertson Trophy Air Race (also known as the London to Melbourne Air Race), which took place in October 1934. The DC-2 *Uiver*—"Stork" in Dutch —registration PH-AJU, flown by K. D. Parmentier and J. J. Moll for KLM, placed second. (The NASM's Boeing 247-D, flown by the celebrated racing pilot Roscoe Turner, placed third.) Remarkably, the KLM DC-2 was flying a regular passenger route with passengers during the race. The strong showing of two commercial aircraft in a long-distance air race proved the promise of modern airliners for fast, reliable, and safe service on regularly scheduled routes. It also indicated that Douglas Aircraft was the world's premiere commercial aircraft manufacturer.

The origins of the immortal DC-3 lay in a request made by C. R. Smith, the president of American Airlines, for a sleeper transport to replace the bulky and slow twin-engine Curtiss Condor biplanes on its southern transcontinental route. The resulting DST (Douglas Sleeper Transport) was a substantial redesign of the DC-2. It was longer by 30 1/4 inches and wider by 26 inches than the DC-2, giving it a lengthened and more rounded fuselage. The sleeper version had fourteen seats (double wide) that could be converted to seven lower sleeper berths and seven upper berths that could be folded into the ceiling. Thus, the DST could accommodate fourteen passengers in its night-transport configuration or up to twenty-eight in its day-transport version.

The day-transport became known as the DC-3. Instead of sleeping berths, it had twenty-one seats. It was of all-metal construction with cantilevered wings, retractable landing gear, and trailing-edge flaps (lift devices designed to maximize the wing configuration on takeoffs and landings). Instrumentation included an "automatic pilot," or electronic navigational control system. It was powered by twin Wright SGR-1820 1,000-horsepower radial engines with cowlings (streamlined metal covers) designed for aerodynamic efficiency and improved cooling.

The DC-3, piloted by Carl Cover, made its maiden flight on December 17, 1935, after which it became an immediate success—it was not only safe and reliable but the first unsubsidized commercial airliner to turn a profit solely by carrying passengers. By 1938, 95 percent of all US commercial airline traffic was on DC-3s. Moreover, the DC-3 was responsible for increasing the number of passenger miles flown in the US more than five-fold (from 267 million in 1935 when the DST-DC-3 was introduced to 1.369 billion in 1941).

Another reason for the DC-3's success was the number of orders that came in from airlines in foreign countries. KLM, in association with KNILM (Royal Dutch Indies Airways), was the first to acquire and operate the DC-3 on the Amsterdam–Batavia (now Jakarta, Indonesia) route. Soon, airlines such as Air France, Sabena, and Swissair followed suit. By 1938, thirty foreign airlines were flying the aircraft; the following year, 90 percent of the world's airline traffic was carried by DC-3s.

As World War II approached, the DC-3 was pressed into wartime service by the US Army Air Corps—first as the C-41A, a standard DC-3 with military instrumentation and communications equipment, and later as the C-47-DL Skytrain, the first completely militarized version of the DST-DC-3 to be used for troop and cargo transport. (Later USAAF designations were the C-53 and C-117. The US Navy designation for the aircraft was R4D.) The C-47 was affectionately nicknamed the "Gooney Bird," "Doug," and "Dumbo" by Americans. The British called it the "Dakota" or "Dak."

Military versions were also put into operation in the European and Pacific theaters by various foreign militaries, including the Royal Air Force, Royal Canadian Air Force, and Royal Australian Air Force, among others. In Russia, a number of license-built DC-3s (designation PS-84/Li-2) flew during World War II. In addition, Japanese license-built DC-3s (designation L2D) were used widely as transports by the Japanese Navy. In fact, the DC-3/C-47 was so respected that the Nazis seized them from Ceskoslovenské Státní Aerolinie (CSA), the Czech national airline, after the invasion of Czechoslovakia in March 1939. With swastikas prominently displayed on the vertical tail section, these were subsequently flown by Deutsche Luft Hansa.

After World War II, the DC-3/C-47 labored into the Cold War era and beyond. Remanufactured versions were sold at low cost to a host of new airlines that started flying after the war in the United States and other countries. C-47s initiated the Berlin Airlift (Operation Vittles) in 1948 but were considered too small (capacity 3.5 tons) to carry the provisions necessary to feed the citizens of Berlin without making at least one thousand flights per day. The DC-3/C-47 was also used as a gunship (designated AC-47 and nicknamed "Puff the Magic Dragon") by the US Air Force during the Vietnam War. Many DC-3s were converted by having their original piston-powered engines replaced by turboprop engines and five-bladed propellers.

Estimates place production of the DC-3/C-47 in the United States at 10,655 aircraft, the Japanese L2D at 487, and the Russian PS-84/Li2 at 4,937, bringing the world total to 16,079. Despite their age, some four hundred Douglas DC-3s are still flying around the world. This model's remarkable combination of strength, durability, and efficiency made it perhaps the most important airliner in history.

Above: The military version of the DC-3 was known as the C-47 Skytrain.

Opposite: Douglas produced an estimated 10,000 C-47 Skytrains in their Southern California and Oklahoma City plants.

DOUGLAS SBD-6 DAUNTLESS

CHAPTER 11

BY CHRISTOPHER T. MOORE

The Battle of the Coral Sea was entering its second day when scout aircraft from the Japanese and American fleets spotted each other's carriers around 8:20 in the morning on May 8, 1942. The day before, aircraft from the carriers USS *Lexington* and *Yorktown* had sunk the Japanese light carrier *Shōhō* but had not been able to engage the larger Japanese fleet carriers. With the sightings on May 8, the Japanese and American carriers launched their strike aircraft.

Lieutenant Junior Grade William E. Hall was not among those pilots assigned to strike the enemy carriers. He and some of his fellow *Lexington* pilots, along with an equal number from *Yorktown*, would stay with the US carriers and provide protection should a Japanese strike threaten them—and shortly after 11:00 a.m., the Japanese attacked. In the melee that followed, Lieutenant Junior Grade Hall destroyed a Japanese B5N "Kate" torpedo bomber and fought to a draw several A6M Zero fighters despite being severely wounded by a 20-millimeter shell. Hall was awarded the Medal of Honor for his actions that day. The surprising thing about Hall's performance was that he was not flying the American naval frontline fighter, the F4F Wildcat, but an aircraft designed as a dive-bomber: the Douglas SBD Dauntless.

This SBD-5 of VS-51 was photographed in May 1944. NASM's SBD was restored in the markings of this aircraft in 1975.

1934

Ed Heinemann designs Northrop XBT-1

1937

Douglas Aircraft acquires Northrop; XBT becomes SBD

1939

April: US Navy orders 144 SBDs

1940

June: First SBD-1s delivered to US Marine Corps

November: First US Navy SBD-2s delivered to USS *Lexington*

1941

December 10: SBD sinks Japanese submarine *I-70*

1942

June 3-7: SBDs sink four Japanese aircraft carriers and one heavy cruiser during Battle of Midway

1944

June: SBDs attack Guam in type's last mission for US Navy

1948

June 30: Last SBD retired from US Navy

Dive-bombing had a long history in naval aviation, with the US Marine Corps experimenting with the technique as early as 1919. With the advent of aircraft carriers, the navy realized that ship-based aircraft would be limited in size, and bombers could not carry the same bomb load as their land-based counterparts. It sought a method of delivering fewer bombs more accurately, and dive-bombing appeared to be the answer. Experiments were conducted, and official recognition came with the inclusion of dive-bombing in the fleet exercises of 1926. Tests with moving targets followed and helped prove the technique's effectiveness. The 1931 adoption of the bomb fork, or crutch, overcame the major obstacle of possible damage to the aircraft by swinging the bomb clear on release.

The Dauntless was first conceived as the XBT-1 by Northrop Corporation's chief engineer, Edward Heinemann, as the company's entry in the US Navy Bureau of Aeronautics' design competition for a new dive-bomber in 1934. The competition resulted from the navy's desire to transition from its fleet of fabric-covered biplanes, including dive-bombers, to all-metal monoplane types. The Northrop entry bore a family resemblance to the firm's Alpha series of mail aircraft. Heinemann's racy low-wing design also incorporated many of the revolutionary construction techniques used in the earlier aircraft. In an effort to save weight, however, unlike many other carrier aircraft the XBT-1 did not include folding wings. In consultation with the National Advisory Committee for Aeronautics (NACA), Heinemann perforated the split dive flaps on the trailing edge of the wing, allowing air to pass through them. This eliminated tail buffeting and permitted a steeper diving angle, which increased accuracy. The navy approved the design and ordered fifty-four production models, designated BT-1s.

The BT-1, however, had stability problems and, with an 825-horsepower Pratt & Whitney engine, was underpowered. A second prototype, the XBT-2, incorporated the new 1,000-horsepower Wright R-1820-32 engine, which boosted the top speed by 35 miles per hour. The addition of a modified rudder corrected another problem of poor lateral stability. Fully retractable landing gear was also included. The changes resulted in improved stability and better low-speed control.

Jack Northrop had formed his company in 1932 in partnership with Donald Douglas, the majority stockholder. In 1937, due to labor problems, Douglas bought out Northrop's share and then dissolved the company, renaming the factory the El Segundo Division of Douglas Aircraft. Many Northrop employees, including engineer Heinemann, moved to Douglas with the XBT project. Coinciding with the switch in manufacturers, the XBT-2 became the XSBD-1 (for Experimental Scout-Bomber Douglas -1), adding the scout role to the aircraft's functions. After some more minor modifications, the navy ordered 144 SBDs in April 1939. Like other Douglas aircraft the SBD was christened with a "D" name: the Dauntless.

Despite its order, the navy did not consider the SBD-1 fully combat ready. Inadequate fuel capacity was the main problem—the amount of time spent forming up and landing on a carrier made fuel capacity critical. Douglas agreed to address the problem, beginning with the fifty-eighth production model. The navy agreed to accept the first fifty-seven SBD-1s without modification, deciding that the marines, who mostly operated from land bases, could use these aircraft. The marines, therefore, received the first Dauntlesses in June 1940.

The remaining eighty-seven aircraft of the original contract were delivered as SBD-2s. The SBD-1's two small 15-gallon auxiliary fuselage fuel tanks were replaced with two 65-gallon tanks in the outer wings, increasing fuel capacity from 210 to 310 gallons and range increased to 1,200 miles. The SBD-2 also had an autopilot for long overwater flights. While these modifications improved range, the increased weight hurt performance; often, one of the two .50-caliber fuselage-mounted guns was removed to compensate. Deliveries of the SBD-2 to the USS *Lexington*'s VS-2 scouting and VB-2 bombing squadrons began in November 1940, with USS *Enterprise*'s VS-6 and VB-6 following shortly.

The two initial Dauntless models saw the first combat in the Pacific during the attack on Pearl Harbor on December 7, 1941. Marine Air Group (MAG) 11, equipped with the SBD-1, was caught on the ground, and all aircraft were either damaged or destroyed. Eighteen Navy SBD-2s launched from the *Enterprise*, which was returning from Wake Island, arrived just as the Japanese were attacking. Seven Dauntlesses were shot down or crash-landed. Two Japanese aircraft were claimed shot down by Dauntless pilots. Three days later, the SBD gained the distinction of destroying the first Japanese warship of World War II when Lieutenant Dickinson of VS-6 sank the Japanese submarine *I-70* off of Hawaii.

Deliveries of the new SBD-3 began in March 1941 and were stepped up after the attack on Pearl Harbor. This was the main variant used in the major battles of 1942. The SBD-3 brought the Dauntless up to full combat standards. Self-sealing wing tanks, crew armor, and an armored windscreen were all introduced.

After the attack on Pearl Harbor, American operations took the form of hit-and-run raids by the carriers *Enterprise*, *Lexington*, and *Yorktown* against remote Japanese positions in the spring of 1942. Dauntlesses were used heavily to attack ships and shore installations. One of the plane's lesser-known exploits during this period occurred during the famous Doolittle Raid: it was an SBD that spotted a Japanese picket boat ahead of the task force, precipitating the B-25s' early launch.

Above: The first production model Dauntless accepted by the Navy, the SBD-1, was not considered suitable for shipboard service and was assigned to the Marine Corps. Here marine SBD-1s stationed at Quantico, Virginia, fly in formation, circa 1941.

Opposite: The National Air and Space Museum's SBD-6 Dauntless (Bureau Number 54605) was perhaps the last of its type in active service. It was earmarked for the national collection when it was stricken in 1948. The Smithsonian accessioned it in 1961 and it was restored in 1975 for the opening of the museum on the National Mall a year later.

A Douglas SBD-3 Dauntless from
Scouting Squadron 5 (VS-5) assigned to the
USS Yorktown (CV-5), circa 1941. The first
Yorktown was sunk at the Battle of Midway.

In May 1942, the United States faced its first major operation against the Japanese fleet
at the Battle of the Coral Sea, which resulted when Adm. Chester Nimitz sent the *Yorktown*
and *Lexington* to block the Japanese thrust toward Australia. What followed was the
world's first carrier duel, in which the opposing ships never came within sight of each other.
Despite the fact that the Japanese sank the larger US fleet carrier *Lexington*, the battle
was a strategic victory for the Americans because it was the first time a Japanese advance
had been halted. During the battle, Dauntless aircrew Lieutenant Junior Grade Hall of
Lexington's VS-2 and his gunner, Seaman 1st Class John Moore; VS-2 squadron mates Ensigns
John Leppla and Radioman John Liska; and Lt. Stanley "Swede" Vejtasa and Radioman
Frank B. Wood of *Yorktown*'s VS-5 achieved remarkable results against much faster Japanese
fighters. Vejtasa, who went on to a successful career as a fighter pilot, commented that with
a more powerful engine the SBD would have made a good fighter aircraft. In less than a
month's time, Dauntlesses would improve on their Coral Sea performance.

Experience gained in the Battle of the Coral Sea prompted the addition of twin
.30-caliber machine guns in the rear cockpit, increasing firepower. The added weight of
these improvements was offset to a certain extent by the use of Alclad, a corrosion-resistant
aluminum, to replace the dural skin of the earlier models and by the removal of the flotation
equipment that was standard on the SBD-2. The service ceiling improved from 26,000 to
27,100 feet, but maximum speed fell by a small margin to 250 miles per hour, earning the
model the tongue-in-cheek nickname "the Speedy Three."

With the failure of the southern expedition, Japanese planners looked toward the
central Pacific and the US base on Midway Island. The plan was to gain a base from which to
threaten the Hawaiian Islands and thus draw the remaining US carriers out to be destroyed
in a major fleet engagement. The three US carriers that took part in the battle carried 112
Dauntlesses. Most were the latest model, but a few SBD-1s and SBD-2s were also aboard.
While the Japanese had a much larger fleet, the sides were more evenly matched in the
crucial area of airpower. The Japanese had four aircraft carriers; the United States had three
carriers plus land-based aircraft at Midway, including nineteen SBD-2s.

On June 4, 1942, after the Japanese opened the battle with a strike on Midway Island, the American carriers launched their aircraft. Because of the uncoordinated nature of the attack, the SBD squadrons had trouble finding the enemy carriers; the USS *Hornet*'s SBDs never located them. Lieutenant Commander Wade McClusky, commander of the *Enterprise*'s Air Group, however, made a navigational guess that brought his VS-6 and VB-6 right over the Japanese carriers. Previous attacks had drawn down the Japanese fighter screen, and the Dauntlesses found the targets wide open. Japanese indecision as to launching further strikes on the island or attacking the recently discovered American carriers left bombs and torpedoes, along with aviation fuel, scattered on their carriers' decks. At the same time McClusky's group attacked, VB-3 from *Yorktown* arrived. The combined onslaught rained thirty-nine bombs on three Japanese carriers in three to four minutes, and eleven direct hits mortally damaged the *Akagi*, *Kaga*, and *Sōryū*. The fourth carrier, *Hiryū*, was located later and also sunk by Dauntlesses. Japan lost four carriers and many of its experienced aviators compared to thirty-five Dauntlesses from the six navy and one marine SBD squadrons engaged. The SBD deserves much of the credit for stopping the Japanese and allowing the United States to gain an equal footing in the Pacific.

The Dauntless also played a significant role in the first major American ground offensive: the fight for Guadalcanal. Marine SBDs based on the island attacked Japanese ships—known as the "Tokyo Express"—attempting to reinforce the island. Ship-based SBDs also participated in the eastern Solomons campaign, of which Guadalcanal was a part, and sank another Japanese carrier.

While the SBD is most often associated with the Pacific theater of operations, it did serve in a limited capacity in the Atlantic. In November 1942, Dauntlesses flew from the carrier *Ranger* and the escort carriers *Sangamon* and *Santee* in support of Operation Torch,

On April 25, 1944, USMC Maj. Gen. James Moore, Commanding General of the 1st Marine Aircraft Wing and Emirau Island in the Bismarck Archipelago, greets the first marine SBD-5 pilots to land safely on the island. The Dauntless began and ended its service in marine squadrons.

the invasion of North Africa. In contrast with the naval actions of the Pacific, SBD attacks during Torch were mostly against ground targets in support of the Allied landings, although some Vichy French ships were also attacked.

SBDs from the *Santee* also conducted antisubmarine patrols in the Atlantic during 1943, but the TBM Avenger was better suited and used more often for this work. Marine Dauntlesses operated from the Virgin Islands in the patrol and scouting role until mid-1944. The last offensive mission for the Dauntless in the Atlantic was an attack on enemy shipping in Norway in which SBDs from the *Ranger* attacked several ships in Bodø harbor and roadstead. They sank two ships, shared in the destruction of two more, and damaged a further two.

Even before the war started, the navy had earmarked the new Curtiss SB2C

Helldiver as a replacement for the Dauntless. However, development problems plagued the Helldiver, and it was not ready for combat in the summer of 1943 when the new *Essex*-class carriers that it was supposed to outfit were launched. The Dauntless soldiered on, but its role on the new carriers changed. The new *Essex* carriers could carry one hundred airplanes, compared to eighty on the older carriers. Scout squadrons were eliminated and their duties taken over by an increased number of Hellcats and Avengers, both of which had long-range capabilities. From then on, Dauntlesses would serve almost exclusively as strike aircraft.

SBDs continued to fly throughout 1943, and not until late in the year did the SB2C Helldiver finally enter service. Even then, many navy pilots did not see it as much of an improvement. They preferred the SBD's more responsive controls, which made it an easy plane to fly when lightly loaded. In addition, the SBD had none of the Helldiver's adverse stall characteristics or marginal handling at low speeds, and it required fewer maintenance hours than did the Curtiss.

In spite of its shrinking role, Douglas continued to modify the Dauntless throughout the war in an attempt to improve performance. The SBD-4 was introduced in late 1942 with an improved 24-volt electrical system that allowed for the installation of radar. It also featured a Hamilton-Standard Hydromatic propeller. But at 245 miles per hour maximum speed, this was also the slowest version.

Bottom left: The SBD-6's rear-position crewman sat in a swiveling seat. Facing forward he could operate the radio; facing to the rear he fired the aircraft's .30-caliber machine guns.

Bottom right: The cockpit of the National Air and Space Museum's SBD-6 Dauntless includes the reflector sight (located above the instrument panel). This was introduced on the SBD-5 and replaced the less accurate old-style tubular telescopic sight.

Early in 1943 the SBD-5, with a larger engine, started to enter squadron service. Two additions that increased bombing accuracy were the replacement of the early models' telescopic sight with a reflector sight and the installation of a heated windscreen that eliminated the fogging that had plagued dive runs. Radar was also more commonly seen on this model, but the additional weight of extra equipment largely cancelled out the increased horsepower. The SBD-5 was the most produced variant of the Dauntless and served throughout the battles of 1943.

The last Dauntless, the SBD-6, featured a still-larger engine, which raised top speed to 262 miles per hour, and a service ceiling to 28,600 feet. In all other ways, however, it differed little from the previous model. By the time it entered production, the Helldiver was replacing the Dauntless in the fleet, and the SBD-6 remained, for the most part, stateside. This was the case with the NASM's Dauntless, Bureau Number (BuNo) 54605, which was the sixth SBD-6 model produced—it was accepted by the US Navy on March 30, 1944, but spent its entire operational career at Naval Air Station, Patuxent River, Maryland. By the end of production, Douglas had produced 5,936 SBDs, second only to the C-47/R4D as the most numerous of the company's many aircraft types.

The Dauntless was also used by the US Army. The success of German dive-bombers during the early years of the war in Europe convinced some army leaders of the need for a US version. Limited experience with this type, coupled with no time to develop a new design, dictated that the US Army Air Forces (USAAF) place orders for the SBD. In army service it was known as the A-24 Banshee, the main differences being the lack of a tail hook and a larger pneumatic tailwheel. The Banshee was delivered as the A-24, A-24A, and A-24B (equivalent to the SBD-3, SBD-4, and SBD-5, respectively). The idea of dive-bombing was not widely supported in the USAAF, however, and the Banshee was not used extensively.

In addition to the United States, New Zealand (in the Solomons) and the Free French (in Europe) also flew SBDs. As late as 1949, the French were using SBDs to carry out attacks on communist terrorists in Indochina. Mexico also flew Banshees, which it had received for patrols in the Gulf of Mexico during World War II, as border patrol planes until 1959.

Despite the introduction of the Helldiver, Dauntlesses continued in navy service until July 1944, when they participated in their last navy mission during an attack on Guam. The marines continued to use them in the Philippines campaign. By the end of World War II, most Dauntlesses had been relegated to training and utility roles. A few marine SBDs, however, worked at neutralizing bypassed garrisons in the Solomons until the end of the war. The NASM's Dauntless was perhaps the last SBD in service with the navy, being stricken on June 30, 1948.

Considered underpowered and scheduled for replacement before the war began, the Douglas SBD Dauntless lived up to the nickname given to it by its crews: Slow But Deadly. Serving throughout the war, Dauntlesses sank more than 300,000 tons of enemy shipping, including at least eighteen warships, ranging from submarines to battleships. Records also show that the number of enemy aircraft SBDs shot down outnumbered the SBDs that were lost—an unusual feat for a bomber. The SBD was the only American aircraft to fly from ships in all five of history's carrier battles. During 1942, SBDs were the primary weapon in the US war effort in the Pacific, almost single-handedly sinking six enemy carriers. By the time its replacement, the Curtiss SB2C Helldiver, entered service, the Dauntless had seen US naval forces through the most crucial part of the conflict, making them one of the most important American aircraft in the theater.

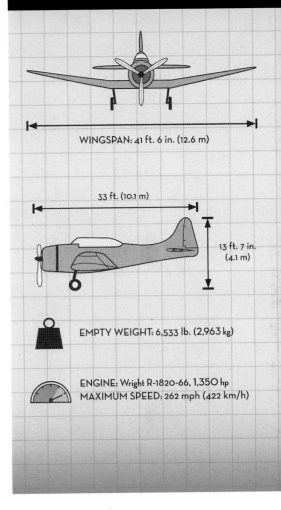

SPECIFICATIONS
DOUGLAS SBD-6 DAUNTLESS

WINGSPAN: 41 ft. 6 in. (12.6 m)

33 ft. (10.1 m)

13 ft. 7 in. (4.1 m)

EMPTY WEIGHT: 6,533 lb. (2,963 kg)

ENGINE: Wright R-1820-66, 1,350 hp
MAXIMUM SPEED: 262 mph (422 km/h)

NORTH AMERICAN P-51D MUSTANG

CHAPTER 12

BY JEREMY R. KINNEY

On August 6, 1944, Maj. George E. Preddy Jr. led the 352nd Fighter Group of the Eighth Air Force on a long-range escort mission of B-17 Flying Fortress bombers against the Nazi capital, Berlin. Before the group reached the target, thirty Messerschmitt Bf 109 fighters approached their formation from the south. From 27,000 feet, Preddy and two other pilots attacked them from above and behind. Preddy immediately shot down one Bf 109 and then three more. Following the fighters down to 5,000 feet, he destroyed two more. In the course of a few minutes, Preddy had become the first US pilot in the ETO to destroy six aircraft in one combat mission and earned the Distinguished Service Cross. His North American P-51D Mustang, named *Cripes A'mighty 3rd* after his trademark saying when he played dice, gave him the speed, altitude, range, and firepower needed to fight the Luftwaffe in the heart of Germany. Before his death from friendly fire in December 1944, Preddy became the highest-scoring P-51 ace of the war with 26.83 aerial and 5 ground victories.

The high-flying, long-range P-51 Mustang escort fighter was a war-winning weapon for the United States and its allies. As American Mustang pilots protected bombers and pursued their enemies in the air over Europe and the Pacific, they and their airplanes earned a place in the annals of military and aviation history.

The National Air and Space Museum's P-51D Mustang is painted in the colors and markings of the 351st Fighter Squadron, 353rd Fighter Group, Eighth Air Force, a unit typical of the many that escorted bombers deep into Germany, and named *Willit Run?*.

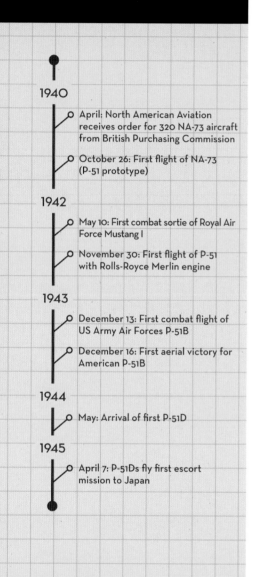

1940

April: North American Aviation receives order for 320 NA-73 aircraft from British Purchasing Commission

October 26: First flight of NA-73 (P-51 prototype)

1942

May 10: First combat sortie of Royal Air Force Mustang I

November 30: First flight of P-51 with Rolls-Royce Merlin engine

1943

December 13: First combat flight of US Army Air Forces P-51B

December 16: First aerial victory for American P-51B

1944

May: Arrival of first P-51D

1945

April 7: P-51Ds fly first escort mission to Japan

The Mustang was the product of an international aeronautical heritage dating back to the early 1930s. North American Aviation was an outgrowth of the General Aviation Corporation that became a major aircraft manufacturer under the leadership of former Douglas Aircraft employee James H. "Dutch" Kindelberger, who moved the company from Dundalk, Maryland, to Los Angeles, California, in 1935. There, the company soon became known for its two-seat high-performance training airplanes, beginning with the NA-16 that air forces in North and South America, Europe, and Asia purchased in large quantities. Those contracts positioned North American for more opportunities in the world military aviation market.

Kindelberger found those opportunities for the small company as military and industrial mobilization reached unprecedented levels in Europe in 1940. When the Anglo-French Purchasing Commission came to the United States shopping for military aircraft, Kindelberger and his team rejected the opportunity to manufacture the Curtiss P-40 fighter in favor of producing an all-new design in less than three months with the same liquid-cooled 1,300-horsepower Allison V-1710 V-12 engine. The P-40 was America's frontline pursuit airplane, but it offered poor speed, range, and altitude performance compared to European and Asian designs. The head of procurement for the army air corps, Col. Oliver P. Echols, who was in favor of a new fighter, facilitated the official arrangement between North American and the Europeans. In April 1940, a contract was signed for four hundred aircraft, designated the NA-73. The first complete airplane was ready 120 days later and flew on October 26, 1940.

North American's chief engineer, Edgar Schmued, an Austrian national who grew up in Germany, led the NA-73 design team and incorporated several new innovations. To make the monoplane even more streamlined, the team placed the air intake and the engine radiator on the underside of the fuselage, making the internal flow of air so efficient it actually provided a small measure of forward thrust to the airplane, called the Meredith effect. The Mustang wing featured a laminar-flow airfoil profile that generated the lowest amount of drag achieved by any aircraft up to that time, making the air move smoothly over the top of the wing in evenly compressed layers while producing maximum lift. For production, the use

of straight wingtips and existing components from North American's AT-6 trainer made the fighter easy to manufacture in large numbers.

The first flight of the production NA-73 took place in April 1941. The series of tests that followed revealed the design's outstanding performance, especially in level-flight speed and low-altitude operations. The British Royal Air Force (RAF) named the new aircraft the Mustang Mk I and used it for tactical photo reconnaissance. The US Army Air Corps placed an order for the P-51A Mustang fighter and a dedicated dive-bomber version called the A-36 Apache. Mustangs Is became operational in August 1942 in Europe; A-36s went into combat in North Africa in June 1943.

The Mustang I, P-51A, and A-36 used the same Allison V-12 engine as the P-40, which excelled at low and medium altitudes and was acceptable for ground attack and dive-bombing but was ill-suited for high-altitude performance, which was a critical in fighter aircraft. Looking for more performance, early in 1942 the RAF retrofitted four Mustangs with 1,650-horsepower Rolls-Royce Merlin engines equipped with two-stage superchargers and optimized to produce sea-level horsepower up to 30,000 feet. British and American fighter pilots who flew the converted Mustangs provided glowing reports on their performance, including a speed increase of 50 miles per hour. Back in California, North American installed Merlins in two aircraft, designated XP-51Bs, and started flight evaluations in November 1942. Overall, the Merlin made the Mustang the fastest and highest-flying piston fighter of the war at speeds approaching 450 miles per hour and altitudes up to 40,000 feet.

North American would build more than fourteen thousand Mustangs at its factories in Inglewood, California, and Dallas, Texas. The average unit cost was $50,895. Unable to keep up with production demands from both the British and American governments, Rolls-Royce

Above: P-51C Mustangs await final preparations before initial test flights at the Texas Division of North American Aviation outside Dallas in 1944.

Opposite: This early P-51 was photographed in flight around 1943. The four 20-millimeter cannons have been airbrushed out to make it appear to be the follow-on P-51A.

licensed the Packard Motor Car Company to produce over fifty-five thousand Merlin engines in the United States. The Packard Merlin powered the Mustang exclusively.

The US Army Air Forces used Mustangs in all theaters of World War II for bomber escort, ground attack and strafing, and photo reconnaissance. In Europe, the strategic Eighth and Fifteenth Air Forces and the tactical Ninth Air Force used the greatest number of Mustangs with twenty-one operational groups. The Fourteenth Air Force in China and the Tenth Air Force in India operated three and two groups, respectively. In the Pacific, the army air forces employed six Mustang groups. The Fifth Air Force in Okinawa had three. The rest were stationed on the island of Iwo Jima in support of the Twentieth Air Force's strategic bombing campaign against Imperial Japan. After the change to night-bombing tactics, the Mustangs of the Twentieth conducted the first land-based fighter attack on Tokyo in April 1945.

The Mustang became a legend as an escort fighter in the air war over Europe. The Eighth Air Force stationed in England conducted its first bombing raid on targets in German-occupied France beginning with Rouen in August 1942. Over one hundred B-17s and B-24s bombed the town of Lille the following October. These raids allowed the Eighth Air Force's bomber units to gain operational experience under the protection of Republic P-47 Thunderbolt and Lockheed P-38 Lightning fighters, which were introduced during the summer of 1943. The limited range of these two fighters, both of which nonetheless became famous over the course of the war, prevented them from venturing far into German territory.

As the Eighth's bombardment groups ventured beyond the range of escort fighters, the price of daylight operations became horribly apparent. American air leaders believed that heavily armed B-17s and B-24s flying at high altitude in defensive "combat box" formations could protect themselves from German Luftwaffe fighters. Raids on important German targets proved otherwise. During one week in October 1943, the Eighth lost 150 bombers to German fighters and antiaircraft artillery, or "flak." Each bomber carried a ten-person crew, which meant an overall loss of 1,500 men killed or captured. Lieutenant General Ira C. Eaker, commander of the Eighth Air Force, ordered a halt to all attacks beyond the range of escort fighters. The Luftwaffe had retained air superiority over Germany and with it the ability to continue producing war materiel.

Sergeant Conige Mormon prepares fellow Tuskegee Airman Lt. Clarence "Lucky" Lester's P-51C, *Miss Pelt*, for combat at their base in Italy in 1944.

The Eighth needed a fighter capable of escorting bombers deep into German territory. North American fitted a P-51B with an extra internal 85-gallon tank in August 1943. The added weight affected longitudinal stability and performance until the gasoline was used, but that addition, combined with external drop tanks that could be jettisoned when needed, extended the range of the Mustang to 850 miles—enough to fly to Berlin and back. The combat introduction of the P-51B Mustang on December 1, 1943, allowed the Eighth Air Force to renew deep-penetration missions. The B model could outperform any German fighter and destroy them with its four wing-mounted .50-caliber machine guns. The

Mustangs of the Eighth flew their first escort mission to Kiel in northern Germany on the Baltic Sea on December 13.

The new commander of the Eighth Air Force, Lt. Gen. James "Jimmy" Doolittle, along with the VIII Fighter Command's Maj. Gen. William E. Kepner (see also *Explorer II* chapter), devised new tactics reflecting the capabilities of the Mustang. In addition to their escort duty, the P-51 squadrons were to seek out and destroy the Luftwaffe's fighter force. They were to meet the Luftwaffe in the air and on the ground through "fighter sweeps" of enemy airfields. The introduction of the improved K-14 computing gunsight and the creation of "Clobber Colleges" to provide additional fighter pilot training made Mustang pilots even deadlier. The numerical superiority and performance of the P-51, combined with offensive tactics and proficient and aggressive pilots, assured that the army air forces would prevail against the Luftwaffe.

With the Mustang, the Eighth and Fifteenth Air Forces renewed the bomber offensive in German-occupied Europe. "Big Week" of February 1944 witnessed over one thousand American bombers acting in concert with the RAF and beginning a two-month assault on German industry and the Luftwaffe. The Nazis lost eight hundred fighters and, more importantly, many experienced pilots. Quickly, the Luftwaffe lost its previously unchallenged control of the air over Europe. March 1944 witnessed the first fully escorted Berlin raid. Bombers of the Eighth Air Force experienced the lowest loss rates of the war in May 1944. Allied air superiority over occupied Europe facilitated the Normandy invasion the following June.

The next variant of the Mustang, the P-51D, appeared in the summer of 1944. It featured three important improvements on previous versions: a bubble canopy and cut-down rear

Ace Lt. Vernon L. Richards of the 361st Fighter Group is at the controls of his P-51D Mustang, *Tika IV*, somewhere over England in 1944.

The P-51 Mustang's forward profile was similar to the German Messerschmitt Bf 109, making recognition difficult for a period after its introduction into combat.

upper fuselage that increased overall visibility; six rather than four .50-caliber machine guns; and almost two thousand rounds of ammunition. The D became the most widely produced variant of the Mustang, with approximately eight thousand leaving North American factories.

The P-51 fighter groups in Europe claimed almost five thousand enemy aircraft shot down in the air and four thousand destroyed on the ground during the war. Between February 1944 and April 1945, the 357th Fighter Group of the Eighth became the highest-scoring Mustang unit with 595 aerial and 106 ground victories. The first African American combat pilots in US history made up the 332nd Fighter Group, called the "Red Tails" for the distinctive recognition markings on their Mustangs. Despite cultural and institutional resistance to these pilots' participation in the war, their professionalism and dedication helped protect the bombers of the Fifteenth Air Force in Italy.

The National Air and Space Museum's P-51D-30-NA, serial number 44-74939, represents one of the P-51 groups that fought in Europe. It entered the army air forces inventory in July 1945 and never saw combat, serving in units assigned to Andrews Army Air Field outside Washington, D.C., and later to Freeman Army Air Field in Indiana. It featured "Guard The Victory, Join the AAF" painted in large black letters on each side of the fuselage under the cockpit and was probably flown during recruiting drives in the immediate wake of World War II. The Mustang accrued eleven months and 211 flying hours of operational service before being set aside for its eventual transfer to the Smithsonian Institution in 1949.

Smithsonian curators selected the Mustang for inclusion in the World War II aviation exhibition at the National Air and Space Museum, painting the artifact in the markings of the 351st Fighter Squadron of 353rd Fighter Group assigned to the Eighth Air Force. The 353rd transitioned from P-47 Thunderbolts to Mustangs in September 1944 and flew bomber-escort and strafing missions for the remainder of the war in Europe. It claimed 330 aircraft shot down and 414 destroyed on the ground and received the Distinguished Unit Citation for its support of the airborne landings in Holland.

As Mustang groups waged war against Nazi Germany and Imperial Japan, many of their pilots became aces and heroes in the process. After George Preddy, Lt. Col. John C. Meyer and Capt. Don Gentile were top Mustang aces, with twenty-four and twenty-three victories respectively. Major James H. Howard, a former naval aviator and American Volunteer Group pilot, became the only fighter pilot in Europe to receive the Medal of Honor. On January 11, 1944, he and his P-51B *Ding Hao!* (American slang for the Chinese phrase meaning "number one") singlehandedly attacked thirty German fighters attacking the B-17s of the 401st Bombardment Group near Berlin. He shot down four in a continuous thirty-minute battle and continued to break up attacks after he ran out of ammunition without losing a single bomber.

The Mustang found a place in the post–World War II American military establishment as well. It was no longer used as a frontline fighter in later conflicts but served as a secondary weapon as newer and faster jet aircraft, such as the Lockheed P-80 Shooting Star, came into service. They remained the main piston-engine fighter for the newly independent US Air Force and received the new F-51 designation in 1948. Mustangs formed the core of the air national guard and air force reserve units across the United States.

The air force's Far East Air Forces, the Republic of Korea Air Force, the Royal Australian Air Force, and the South African Air Force flew Mustangs for close air support, ground attack, interdiction, and photo-reconnaissance missions during the Korean War. The planes' long range, rugged construction, maneuverability, and offensive armament of machine guns, bombs, and rockets enabled units flying them to start operations from Japan and continue from the rough airfields of Korea. Unfortunately, one design characteristic of the Mustang made it vulnerable as a fighter-bomber: the entire cooling system ran along the bottom of the aircraft between the engine and the air intake, where a single bullet through the radiator or a coolant pipe was usually enough to down an F-51. The Mustang units suffered considerable losses to ground

These cockpit views of the Smithsonian National Air and Space Museum's North American P-51D, *Willit Run?*, shows what fighter pilots saw as they climbed behind the controls. Mustang pilots operated in incredibly cramped quarters for long periods of time, but a well-designed layout ensured they had all flight, propulsion, and armament controls at their fingertips.

fire before being replaced by North American F-86F Sabre jet fighter bombers in 1953. After Korea, the F-51 remained in air force service until 1957.

P-51s also served in the air forces of other countries during and after World War II. Besides the Royal Air Force, British and Commonwealth operators included Canada, Australia, New Zealand, and South Africa. The Indonesian, Philippine, and both the Nationalist and Communist Chinese air forces flew them. France, Sweden, Switzerland, and Italy operated P-51s in Europe through the 1950s. The Israeli Air Force used them from its formation in 1948 through the Suez Crisis in 1956. Latin American air forces, including the Dominican Republic and Guatemala, kept Mustangs in their inventories until the late 1980s, while El Salvador employed Mustangs in their last combat operations during the Football War against Honduras in July 1969.

The availability of surplus Mustangs and other high-performance fighters (such as the Lockheed P-38 Lightning, Bell P-39 Airacobra, Vought F4U Corsair, and Grumman F8F Bearcat) after World War II and into the 1950s helped create the "warbird" community. Owners of these aircraft used them in a variety of ways, from pure leisure flying to air show demonstrations and motorsports competition. Many civilian Mustang owners, recognizing that they were flying the fastest propeller-driven production aircraft ever made, took them air racing in the post–World War II period. They remembered the widely popular prewar National Air Races at Cleveland, where clever individuals took readily available technologies and built dedicated air racers to achieve fame and fortune in 1930s Depression-era America. The short-lived National Air Races (1946–1949) and the National Championship Air Races (1964–present) employed warbirds, primarily Mustangs, as the main vehicles for long-distance and closed-course competition.

Echoing the longstanding American tradition of technical ingenuity, modification, and "hot-rodding" found in other motorsports, racing teams carried out their visions of improving upon the original Mustang design to make their aircraft faster and lighter. The Unlimited Class of the National Championship Air Races, flown primarily at Reno, Nevada, became the marquee event for this activity as aerodynamic streamlining, structural modification, and increased horsepower turned an aircraft designed as a high-altitude, long-range fighter

Civilian postwar Mustang owners included aviation pioneer Jacqueline "Jackie" Cochran, who used her P-51B to compete in long-distance air races and break speed records.

capable of cruising at 350 miles per hour into a racer intended to fly fast and low at speeds approaching 500 miles per hour over the high desert course. Since 1964, Mustang racers have won twenty-eight Unlimited championships, more than any other type of aircraft.

One war surplus Mustang stands out as a racer, record setter, and contributor to postwar aviation. The famed Hollywood stunt pilot Paul Mantz bought a standard P-51C, named it *Blaze of Noon*, and modified it for long-distance flying. The changes included extra fuel tanks and the conversion of the interior spaces of the wings to carry even more fuel. Mantz won two consecutive Bendix Trophy transcontinental air races from 1946 to 1947 and set a transcontinental speed record in 1947 in the aircraft. Pan American World Airways pilot Charles F. Blair purchased the Mustang from Mantz in 1949 and renamed it *Excalibur III*. He flew the Mustang from New York to London in less than eight hours in January 1951, which is a record that still stands today for propeller-driven, piston-engine

Charles Blair's modified P-51C Mustang, *Excalibur III*, hangs on display at the Smithsonian National Air and Space Museum's Udvar-Hazy Center in Chantilly, Virginia.

aircraft. Blair developed a system of navigation over polar regions that used carefully plotted "sun lines" to offer precision where conventional magnetic compasses failed. To prove his system, Blair flew *Excalibur III* from Bardufoss, Norway, across the North Pole to Fairbanks, Alaska, in a record-setting ten and a half hours on May 29, 1951. The transpolar navigation opened new opportunities for commercial aviation and introduced the sobering reality that nuclear bombers could also follow the same routes during the tense early days of the Cold War. Pan American bought *Excalibur III* from Blair and donated it to the Smithsonian Institution in November 1953.

As Mustangs were phased out of active military service in the late 1950s, newspaper publisher David B. Lindsay Jr., saw a new market opportunity. He purchased surplus aircraft for conversion into long-range high-speed executive aircraft by his company Trans Florida Aviation beginning in 1957. The company sold approximately twenty completely rebuilt Cavalier Mustangs by 1969. As Mustang specialists, Lindsay's company overhauled surplus fighters under contract to the US Air Force and to several Latin American air forces, including the Dominican Republic, Guatemala, Nicaragua, and Bolivia. He purchased the Mustang design from North American and renamed his company Cavalier in 1967. Believing in the promise of the design, Lindsay went on to develop a turboprop-powered version called the Turbo Mustang, which Piper Aircraft purchased and marketed unsuccessfully as the Enforcer counterinsurgency aircraft to small air forces. As the operational lives of these converted Mustangs ended, civilian owners purchased them, brought them back to the United States, and restored them back to their original P-51D configurations.

In the early twenty-first century, approximately three hundred Mustangs survive. Over half are in airworthy condition, flying primarily in the United States and Europe. Many of them have been painstakingly restored with markings that reflect the P-51's important World War II history and appear at air shows featuring vintage military aircraft. The remainder are in the process of being restored, in storage, or displayed in museums around the world.

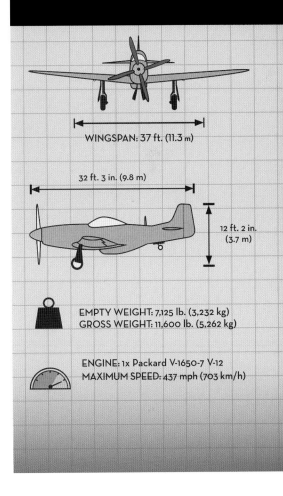

SPECIFICATIONS
NORTH AMERICAN P-51D MUSTANG

WINGSPAN: 37 ft. (11.3 m)

32 ft. 3 in. (9.8 m)

12 ft. 2 in. (3.7 m)

EMPTY WEIGHT: 7,125 lb. (3,232 kg)
GROSS WEIGHT: 11,600 lb. (5,262 kg)

ENGINE: 1x Packard V-1650-7 V-12
MAXIMUM SPEED: 437 mph (703 km/h)

BELL XP-59A AIRACOMET

CHAPTER 13

BY THOMAS J. PAONE

The United States officially entered the jet age on October 2, 1942. On that day, after rapid development under the strictest of secrecy, Bell Aircraft test pilot Bob Stanley officially flew the Bell XP-59A Airacomet from the remote desert sands of Muroc Army Air Field in southern California. The XP-59A neither entered military service nor flew in combat, but it did prove the promise of jet technology to the US military, usher in a new era in military aviation, and allow pilots to receive their first experience with a new type of flight.

Simple in design, the Airacomet was the product of the US attempt to catch up in the field of jet technology. Before World War II, many in the US scientific and military communities were uninterested in jet technology. Resources were limited, and those that were available were dedicated to improving already existing technologies, such as superchargers, not developing new inventions. As the United States entered the war, the situation only worsened, and the US Army Air Forces dedicated all available resources to improving conventional engines and making more aircraft. The situation changed drastically when commander of the US Army Air Forces Maj. Gen. Henry "Hap" Arnold witnessed the first flight of Britain's first jet, the Gloster E.28/39, in April 1941 while meeting with British engineers to discuss jet technology. During that session, Arnold learned that the United States was far behind the rest of the world and needed to catch up.

The first XP-59A is shown on display in the Boeing Milestones of Flight Hall at the Smithsonian National Air and Space Museum. This was the first turbojet aircraft ever designed and flown by the US military.

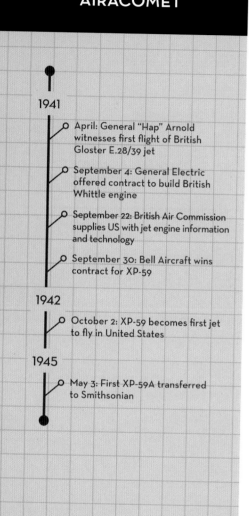

TIMELINE
BELL XP-59A AIRACOMET

1941

○ April: General "Hap" Arnold witnesses first flight of British Gloster E.28/39 jet

○ September 4: General Electric offered contract to build British Whittle engine

○ September 22: British Air Commission supplies US with jet engine information and technology

○ September 30: Bell Aircraft wins contract for XP-59

1942

○ October 2: XP-59 becomes first jet to fly in United States

1945

○ May 3: First XP-59A transferred to Smithsonian

Arnold returned to the United States and immediately briefed members of the Army Air Forces Engineering Division at Wright Field, as well as members of the State Department, on what he had seen. The briefings led to negotiations between the United States and Great Britain, and on September 22, 1941, the British government authorized the British Air Commission to provide their American counterparts all available jet propulsion information, including drawings of the engine; access to primary engineer Frank Whittle; and authorization to build copies of the E.28/39 within American factories, with certain restrictions. The British government even provided a completed W.1X gas turbine jet engine, designed by Whittle, that powered the aircraft. Although Whittle had begun his engine development in the early 1930s, the information was new to many of the American engineers and proved invaluable to the development of the first American jet engine.

While negotiations were underway with the British government, the US Army Air Forces began searching for manufacturers in the United States who could rapidly develop the project under strict security. The project was broken into two main components: the design and construction of the exotic jet engine and the design and construction of the aircraft it would power. Careful consideration had to be taken to ensure that the jet program would not cause delays in other projects for the war effort.

On September 4, 1941, after a personal meeting in the office of Major General Arnold, the General Electric Company (GE) was offered a contract to build jet engines based off on the newest Whittle engine, designated the W.2B, provided by the British government. Brigadier General Oliver P. Echols, who was serving as chief of the Material Division of the US Army Air Forces, recommended GE because the company had experience building turbo superchargers for piston-powered engines, and many of the components of the turbo supercharger were similar to those found in jet engines. GE also had experience with high-temperature metal alloys essential to functional jet engines.

To support the engine portion of the project, Arnold also needed an aircraft manufacturer to work with GE to develop the airframe for the new jet aircraft. On September 5, 1941, Larry Bell and his chief engineer, Harland M. Poyer, met with Echols and Arnold and were briefed about the project. Bell Aircraft was chosen to design the aircraft for a variety of reasons. Foremost, the company was known for innovative designs like the P-39 Airacobra. In addition, it was not overworked by other war-related projects, and its factory in Buffalo, New York, was located near General Electric in Schenectady, allowing the two companies to easily collaborate. The contact was officially awarded on September 30, 1941, with the first airplane scheduled to be delivered eight months later.

Secrecy was paramount from the very first day of development. Arnold had promised the British government that all precautions would be taken

to ensure the information they provided would not be leaked. After Bell Aircraft was awarded the contract, six engineers were personally briefed by Larry Bell and given nothing more than sketches of the yet to be produced jet engine. Those at GE and within the US Army Air Forces at Wright Field were equally secretive about the development, with only a limited number of people at any one time knowing about the project. Both companies adopted names for their individual components that disguised the true nature of the project. Bell Aircraft chose XP-59A for the aircraft component of the project to make it appear to be a modification to a previously proposed XP-59 fighter aircraft (a proposed twin-tail aircraft with a dual-rotation pusher propeller that was in no way similar to the XP-59A jet-powered aircraft). General Electric followed equally strict security protocols when they designated the jet-engine the Type I-A. General Electric was producing several piston engine turbo superchargers that were designated Type A through Type F, and the jet engine was referred to as a turbo supercharger in official documents to hide the true details of the project.

The high levels of secrecy, however, did lead to some problems with the design of the aircraft. Arnold gave orders forbidding the XP-59A airframe from being tested in the Full-Scale Wind Tunnel operated by the National Advisory Committee for Aeronautics (NACA). As a result, engineers at Bell had to rely on less reliable data from a smaller wind tunnel at Wright Field, and that data was collected only after multiple appeals to Arnold. This resulted in serious problems with the design that hindered performance once the aircraft and engine were mated.

Even with the problems caused by the extreme secrecy of the project, development of both components moved forward at an accelerated pace. Once the initial designs for the aircraft were approved, Bell Aircraft moved at once to begin constructing the first airframe. The company leased a building in Buffalo from the Ford Motor Company and quickly moved machinery into place. Security was increased at the building—entrances were guarded at all times, and windows were painted over to prevent anyone from seeing inside. The manufacture of the plane began in January of 1942. The first airframe was produced almost entirely by hand, as the confidential nature of the project prevented the

Above: This view of the Bell XP-59A Airacomet cockpit showing the aircraft's instruments and control column was taken when the aircraft arrived at the Smithsonian Institution in 1945.

Top: The extreme secrecy surrounding the XP-59 project led to extraordinary measures during the testing of the aircraft at Muroc Air Base. The plane and engines were covered with a large tarp and a fake propeller was placed on the nose whenever it was moved around the base.

Opposite: Bell Aircraft's chief test pilot Robert M. Stanley poses in the cockpit during aircraft testing at Muroc.

team from obtaining components from outside companies, and was very similar to others designed by Bell Aircraft for piston-engine-powered aircraft. The lack of time, as well as the ban on outside engineers, prevented the development of any major new design elements.

The team at GE was facing their own problems with the Type I-A engine. They had a copy of the British W.1X engine, as well as incomplete drawings of the W.2B, but they were trying to produce an improved version of the engines. They attempted to test a prototype with improved components on March 18, 1942, less than six months after beginning the initial work. The test failed, but modifications led to a successful test one month later, on April 18, 1942. In June, Frank Whittle arrived in the United States to assist GE with their engine design and spent about two months working with the company to help overcome excessive exhaust gas temperatures. He returned to England in August 1942 after several issues with the engine had been worked out. By that point, the engine problems had delayed completion of Bell's first airframe, and the engineers at GE realized that the engine would never produce the 1,250 pounds of thrust they had originally predicted.

The jet engines developed for the XP-59 program needed constant attention during testing. Maintenance and a lack of spare parts led to lengthy project delays.

Nevertheless, the engine and airframe were now ready for testing, but only in a place where the high levels of secrecy could be maintained. The jet engines, however, could not be kept a secret once testing began.

Muroc Army Air Field was a training base in the high desert of southern California. In September 1942, the numerous pieces of the airframe and engine were transported by rail from New York to the remote desert base, and engineers from GE and Bell immediately set to work reassembling the aircraft. Almost one year to the day after the contract was awarded, the XP-59A was fully assembled and ready for flight testing. Even at the remote desert base, every measure was taken to conceal the aircraft. Whenever it was moved around, it was covered with a piece of canvas, and a fake wooden propeller was placed over the nose to disguise it as a regular propeller-driven aircraft.

On September 26, 1942, the jet engines of the XP-59A were started for the first time. For the next four days, Robert M. Stanley, the chief test pilot for Bell Aircraft, conducted several high-speed taxi runs to learn the aircraft's controls. During some of these runs, the wheels actually lifted off the ground. Then, on October 1, 1942, the first "flights" of the aircraft occurred during additional tests when it reached an altitude of 25 feet the first time, and about 100 feet the second time. Although not the official test flight, the event marked the first flight of an American-built jet-powered aircraft. On October 2, 1942, the aircraft's first official test flights occurred in the presence of US Army Air Forces representatives. During the first flight, Stanley flew the XP-59A to an altitude of 6,000 feet. "Duration of flight 30 minutes," Stanley's notes read, in part. "Throttle was applied promptly and acceleration during take-off appeared quite satisfactory." His second flight of the day reached an altitude of 10,000 feet, and he later noted, "The speed in level flight at 10,000 feet was surprisingly high . . . I had less trouble and fewer mechanical interruptions than on any other prototype I have ever flown."

Top left: The XP-59 was captured at rest in the Southern California desert.

Top right: Although the XP, YP, and P-59 aircraft were never used in combat, they did showcase the future direction of what would become the US Air Force, as illustrated in this recruitment poster.

The first jet fighter accepted by the US military, the Lockheed P-80 Shooting Star, was developed by engineer Clarence "Kelly" Johnson during the same timeframe but completely separate from the XP-59A. The test XP-80 *Lulu-Belle* is also in the collection of the Smithsonian National Air and Space Museum.

The third flight of the day marked another important event in aviation history. For that flight, Stanley turned control of the aircraft over to Col. Lawrence G. Craige, the chief of the Aircraft Project Section at Wright Field. Craige was on hand to observe the test flights, and as a result he became the first American military pilot to fly a jet-powered aircraft. After one additional flight by Stanley that historic day, the XP-59A would not fly again until the end of the month. The engines were replaced with two new I-As, and an observation seat was added to the nose of the aircraft. Flight testing recommenced on October 30, 1942, but continued problems with the engines limited the aircraft to a top speed of 390 miles per hour—slower than many conventional piston-engine fighters of the day—and led to the first XP-59A being nicknamed *Miss Fire*. A lack of readily available replacement parts and other problems with the new technology led to delays in further official testing of the aircraft by the US Army Air Forces.

While the XP-59A was being tested at Muroc, the next configuration of the aircraft, known as the YP-59, was nearing final design. The first YP-59 models arrived at Muroc on June 21 and 22, 1943, but further changes to the aircraft delayed flight testing until August. Setbacks with a new engine for the aircraft, designated the General Electric I-16 turbojet engine, added further delays to the overall flight testing.

Testing of the YP-59 produced some historic moments as well. Ann G. Baumgartner Carl, a member of the Women Airforce Service Pilots (WASPs), was given the opportunity to fly one of the YP-59s, becoming the first woman to fly a jet-powered aircraft. In her memoir, *A WASP Among Eagles*, Carl recounted the experience: "I pushed the engines to a high scream and started down the runway. As advertised, it took a while to get airborne. Settled into the climb, suddenly the engine noise stopped. Had the engines quit already? No, we (the plane and I) were still climbing. Then I realized the jet noise was behind me. . . . As we slid along silently, it was strange to realize I was the only jet up there, perhaps the only jet over the United States that day."

In continued testing of the XP-59 and YP-59 airframes by both the US Army Air Forces and the navy, the performance of the craft never reached what was intended with the project. Even with upgraded engines and an improved design, the YP-59 reached a maximum speed of just 409 miles per hour. In the end, fewer than one hundred production-version P-59 Airacomets were built, most of which were used for training purposes.

Thus, the P-59 did not become the first jet fighter to enter military service for the United States. That honor went to the P-80 Shooting Star, a fighter designed by Lockheed engineer Clarence "Kelly" Johnson around the same time as the XP-59A in a completely separate project. The XP-59A, however, did help pave the way to the future success of jet fighters. The final air force evaluation of the aircraft concluded that "even though a combat airplane did not result from the development of the X and Y P-59A airplanes, it is considered that the development was worthwhile since it proved that the principle of jet propulsion for aircraft was sound and practical."

The P-59 proved effective in allowing American pilots to become familiar with an aircraft technology that would revolutionize aviation, and its importance was recognized even before World War II came to an end. On April 12, 1945, Henry E. Collins, the vice president of Bell Aircraft, sent a letter to Dr. Alexander Wetmore, the Secretary of the Smithsonian Institution, alerting him to the availability of the first XP-59A airframe from the US Army Air Forces. Dr. Wetmore replied on April 18, 1945, that "this experimental plane, which marks an epoch in the history of the development of aviation in America, where the art saw its birth, would be a most fitting gesture and we should be glad indeed to provide a suitable place for its care and exhibition."

By May 3, 1945, the aircraft was officially transferred to the Smithsonian Institution.

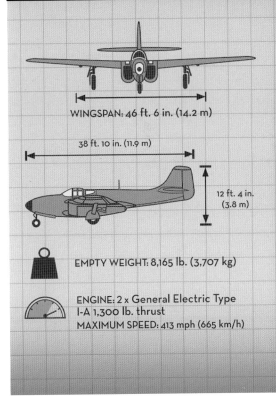

SPECIFICATIONS
BELL XP-59A AIRACOMET

WINGSPAN: 46 ft. 6 in. (14.2 m)

38 ft. 10 in. (11.9 m)

12 ft. 4 in. (3.8 m)

EMPTY WEIGHT: 8,165 lb. (3,707 kg)

ENGINE: 2 x General Electric Type I-A 1,300 lb. thrust
MAXIMUM SPEED: 413 mph (665 km/h)

MESSERSCHMITT ME 262A 1-A SCHWALBE (SWALLOW)

CHAPTER 14

BY EVELYN CRELLIN

The Messerschmitt Me 262 was not only the world's first operational jet—it was a truly formidable combat aircraft, flying faster than its competition and possessing an unmistakable look. Sleek, partially swept wings, turbojet engines, and an arrow shape made the Me 262 an aircraft of both sinister beauty and futuristic allure.

Jet engine development in Germany dated to the mid-1930s. One designer who had worked on developing aircraft based on this new propulsion method was Willy Messerschmitt, who also had been the designer of the Bf 109, Germany's standard fighter during World War II, with more than thirty-three thousand built. Messerschmitt's prototype, designated Me 262 by the German air ministry (the Reichsluftfahrtministerium, or RLM), first flew on April 18, 1941, propelled by a single Junkers piston engine and featuring a conventional tailwheel, since the jet engines were not yet available. Their development proved lengthy and difficult. Completely new ground had to be broken, and there was a shortage of critical materials and testing equipment.

One year later, on July 18, 1942, the Me 262 made its first flight under full jet power, using two Jumo 004 engines. Although the aircraft performed very well, the jet engines remained a problem; they never became truly powerful, each of them

The Smithsonian National Air and Space Museum's Me 262 as exhibited in the Jet Age Gallery. The yellow 7, the aircraft's unit marking, is clearly seen.

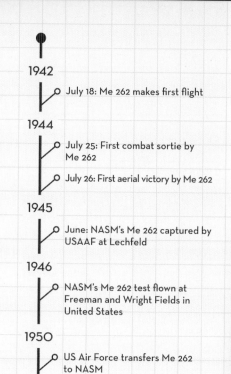

1942

○ July 18: Me 262 makes first flight

1944

○ July 25: First combat sortie by Me 262

○ July 26: First aerial victory by Me 262

1945

○ June: NASM's Me 262 captured by USAAF at Lechfeld

1946

○ NASM's Me 262 test flown at Freeman and Wright Fields in United States

1950

○ US Air Force transfers Me 262 to NASM

providing 1,980 pounds of thrust, insufficient for the Me 262's gross weight of 14,000 pounds. And due to a lack of critical materials (especially chromium, nickel, and molybdenum) needed to harden the blades, the engines had a short lifespan.

Furthermore, the Me 262 had many competitors for Germany's limited resources. There were conventional piston fighters, such as the Bf 109 and single-seat, single-engine Focke-Wulf Fw 190, together considered the backbone of the Luftwaffe's fighter force. Further, the Me 262 was but one of several jet fighter and bomber programs. Messerschmitt's rival, Ernst Heinkel, prompted development of the Heinkel He 280, another twin-engine jet design, and Arado Flugzeugwerke built the world's first jet bomber and jet reconnaissance aircraft, the Ar 234, powered by two Jumo 004 engines and equipped with rocket-assisted takeoff (RATO). There was also the Messerschmitt Me 163 Komet, which became the only rocket-powered fighter/interceptor ever operational. Designed by Alexander Lippisch and equipped with an unreliable rocket engine providing a flight time of about eight minutes, the Me 163 could briefly attain unrivaled speeds of about 560 to 620 miles per hour. Its short flying time, dangerous propulsion system, and high attack speed limited the Komet's wartime success. The National Air and Space Museum has rare examples of all of these aircraft except the He 280.

Following the German defeat at Stalingrad and the rising tempo of the Anglo-American bomber offensive, pressure mounted to produce more capable aircraft for defense of the Reich. The Luftwaffe demanded a sharp increase in jet engine production, and plans were

INTELLIGENCE, (☆☆☆)

made to produce twenty Me 262s per month, starting in January 1944. When Adolf Galland, general of fighter pilots, test-flew an Me 262 on May 22, 1943, he commented that the flight felt like "being pushed by an angel." Galland quickly realized the superior quality and tactical possibilities of the aircraft and argued for one thousand Me 262s to be produced per month. However, there was no chance to expand the production of Me 262—not enough jet engines were manufactured, there was strong competition from other aircraft manufacturers, and the Allies' increasing air control over Germany meant that by early 1944, basically every German aircraft manufacturing site had been repeatedly bombed.

On March 1, 1944, as a result of these attacks, the establishment of *Jägerstab* (fighter staff) created the administrative tool necessary to increase the production of German fighter aircraft despite Allied bombing. The *Jägerstab* and its successor from 1944 on, *Rüstungsstab* (armament staff), decentralized production, relocated it into newly built underground facilities or production sites hidden in forests, and provided all necessary machines, engines, equipment—and labor force. More than 60 percent of laborers in the German aviation industry were unskilled forced laborers, including prisoners of war and concentration-camp inmates. In the actual assembly of aircraft, the number was even higher.

Conditions in the forced labor facilities were horrific. Probably the worst slave labor site in Me 262 production was the Gusen II subcamp near Sankt Georgen, Austria, where an SS-owned company ran the underground facility B8 Bergkristall, producing, among other items, almost one thousand Me 262 fuselages. Inmates called Gusen II "hell of hells," and the average survival expectancy was four months. Up to sixteen thousand inmates worked in the camp at any given time; the official death toll is estimated to be between eight thousand and twenty thousand— most of them victims of Me 262 production.

From the various decentralized manufacturing sites, the assembled subsections were taken to a final assembly plant.

Above: Me 262 FE-111—which would become National Air and Space Museum's Me 262—is displayed among other captured aircraft at Park Ridge, Illinois, in 1945 or 1946.

Opposite: This photo shows an Me 262 assembly plant in southern Germany damaged in a US air raid at the end of World War II. Near the end of the war the desperate Nazi regime employed slave labor in the Me 262 program.

Below: Kill markings on National Air and Space Museum's Me 262. The original markings (left) had been overpainted and clumsily reproduced for an exhibition of captured German aircraft in the United States. A summary of pilot Heinz Arnold's aerial victories were added at right.

THIS PILOT (Recor
1 P. 51.
1 P. 47.
5 B. 17.
42. RUSSIAN PLANES

After workers hung the Jumo 004 engines and conducted an operational test flight, the aircraft was then handed over to Luftwaffe units. In December 1943, an Me 262 test command (*Erprobungskommando*) was established, and in June 1944, a first unit of Luftwaffe pilots started training with Me 262A fighters. Allied pilots saw the Me 262 in action for the first time on July 25, 1944, over Munich, attacking a British de Havilland Mosquito photoreconnaissance aircraft. The first kill of an Allied aircraft by an Me 262 was recorded the following day.

Many Me 262 variants were proposed, prototype-tested, and flown in a variety of capacities—as reconnaissance planes, two-seat trainers, radar-equipped night fighters, and a bomber complete with a glazed nose to accommodate a bombardier. But mostly they were built in two fighter versions: the Me 262A 1-a Schwalbe (Swallow), a fighter/ interceptor equipped with four 30-millimeter machine cannon, and the Me 262A 2-a Sturmvogel (Stormbird), a fighter-bomber equipped with two machine cannons and Wikingerschiff (Viking Ship) bomb racks attached in streamlined fairings under the nose, enabling the aircraft to carry bomb loads up to 1,000 kilograms. Some Me 262s carried additional missile armament, such as the R4M rocket system, which provided twelve 55-millimeter air-to-air missiles under each wing, saturating an area large enough to bring down a B-17.

The Me 262 attacked in a wide, sweeping curve, climbing through Allied formations with an initial speed of 286 miles per hour. With a top speed of 540 miles per hour, it surpassed

US servicemen inspect the engine of an Me 262 after the captured aircraft was brought to the United States in summer 1945.

Museum experts spent countless hours restoring the Me 262's jet engine.

the performance of every other World War II fighter aircraft by more than 100 miles per hour. Thus, its closure rate was extremely fast upon a slow-moving bomber, allowing the pilot only a very short window to aim and fire. And while the battery of four 30-millimeter cannons was devastatingly powerful, the cannon were also slow to fire. Only truly experienced pilots were able to make kills under these circumstances. This combination of fast approach and slow firing rate led to many missed opportunities, saving many an Allied crewman's life.

Beside its unique powerplants, the Me 262 also featured trailing-edge flaps and automatic leading-edge slats that provided stability at low airspeeds. Its 18-degree wing sweep preserved the aircraft's center of gravity while incidentally delaying the onset of compressibility. Each engine's nose held a built-in starter motor to facilitate operations from temporary airfields.

Despite its revolutionary technology, the operational history of the Me 262 was plagued by a variety of technical and logistical problems. Its high fuel consumption allowed only a relatively short flying time. Poor quality of the synthetic material used in its tires frequently led to failures. The aircraft's brakes were unreliable, and its nose gear and undercarriage were notoriously weak and often collapsed, leading to disastrous accidents. Excessive imposed loads during maneuvering often led to the structural failure of the tail plane. The aircraft's need for a 6,000-foot runway also restricted the number of bases from which it could operate, enabling the Allies to target them for bombing. And due to the low speeds

The Me 262's wings are repainted at the National Air and Space Museum's Paul Garber facility during the its restoration.

during takeoff and landing as a result of the engines' slow acceleration to full power, Me 262s were ideal targets for loitering Allied fighters and had to be protected by special squadrons of Fw 190D and Bf 109 piston fighters.

By war's end, 1,433 Me 262s had been produced. Due to a shortage of pilots, fuel, and spare parts, however, never more than fifty or sixty were in operation at any given time, and only about three hundred saw actual combat. While sustaining about one hundred combat losses, Me 262 pilots claimed 542 victories.

The National Air and Space Museum's Me 262, *Werksnummer* (production number) 500491, was built at the end of 1944 at the Messerschmitt Obertraubling plant near Regensburg. It was captured in June 1945 at Lechfeld near Augsburg by a special USAAF team led by Col. Harold M. Watson, who was in charge of the seizure of advanced German aircraft for study in the United States. Watson's team gave the Me 262 the code number 888 and the nicknames "Dennis" and, later, "Ginny H." On June 10, 1945, the aircraft was flown from Lechfeld to Melun, France, for a special aerial display of a group of Me 262s put on for the US Army Air Forces' Gen. Carl A. Spaatz.

Me 262 500491 was then taken to the United States with sixty-five other captured German aircraft aboard the British HMS *Reaper*. At Freeman Field and Wright Field, they received Foreign Equipment (FE) numbers, with the Me 262 that would eventually go to the National Air and Space Museum becoming FE-111. As part of various testing programs, probably early in 1946, FE-111's standard fighter nose was swapped for the reconnaissance nose of another captured Me 262. Once testing was finished, FE-111 was sent to Park Ridge, Illinois, for storage. In 1950, the USAF donated it to the National Air Museum. Restoration began in 1977.

After the aircraft's history was carefully researched, the decision was made to convert it back to its original configuration of a standard Me 262A-1a fighter. Overall, the Me 262 was in a pitiful condition, with corrosion built up in many areas, including the nose, making the conversion back to a standard fighter all the more challenging. Toward the end of the war, German aircraft manufacturers had not built aircraft to last, and corrosion control had been of limited interest. Also, a shortage of aluminum during that time meant much of the Me 262 was built of both aluminum and steel, and wherever these materials touched, electrolytic corrosion resulted that had to be carefully treated. (A shortage of aluminum is also the reason the doors and instrument panel were made of wood.) The cockpit was completely refurbished

to near-new condition, including decals, and components of three Jumo 004 engines were used to rebuild the two powerplants.

A final step was the restoration of the exterior. The plane had been repainted several times, and the original Luftwaffe colors were gone, the original markings obscured. After carefully hand-sanding the aircraft, employees reapplied the original paint scheme, including dozens of maintenance instructions and other markings, as well as the yellow 7 (the aircraft's call number), indicating that the NASM's Me 262 had belonged to III./ *Jagdgeschwader* 7, or III./JG 7 (Group III of the 7th Fighter Wing). Established in southern Germany in fall 1944, this was the largest Me 262 fighter wing in the Luftwaffe. The victory markings of the pilot, Oberfeldwebel (Master Sgt.) Heinz Arnold, were restored as well. Arnold had achieved forty-two victories over Soviet planes in a piston-engine aircraft, and these markings had been transferred to his Me 262. In the Me 262, Arnold had achieved a further seven air victories over Allied aircraft, which were marked with the date and type of aircraft shot down. Arnold, born in 1919, was killed in an air battle in another aircraft in April 1945.

After more than six thousand hours of restoration, the Me 262 was exhibited at the museum's National Mall Building in 1979, where, with its unique history and design and its appearance of a shark-like grin, it has attracted visitors ever since.

Despite its short existence, the Me 262 has given birth to many myths, one of them being Adolf Hitler's alleged role in its deployment. The Luftwaffe was charged with developing

The Me 262's original markings are meticulously restored. White masking tape marks the outline of the aircraft's Balkenkreuz, or "bar cross."

two completely different employment doctrines for the Me 262. However, the *Führerbefehl* (Führer decree) regarding the modification of Me 262 as a *Blitzbomber*, or "revenge bomber," was only in effect between June and November 1944, and only about 30 percent of Me 262 airframes were diverted to be equipped as bombers. Considering the miniscule role the Me 262 played at the end of the war, speculations about Hitler's impact on the employment of the Me 262, and thus possibly the outcome of the war, are without merit.

The controversy, however, gained momentum after 1945 because it apparently served to demonstrate irreconcilable discrepancies between Adolf Hitler and the Luftwaffe leadership. After a fallout with Luftwaffe head Hermann Göring in the last months of the war, Adolf Galland, in an apparent "mutiny," established *Jagdverband* 44 (JV 44, or Fighter Unit 44) equipped with Me 262s armed with the powerful R4M missile system. Often called "the elite of the elite," the unit comprised Germany's most famous aces, many of them with dozens of kills—ten members of JV 44 held the Knight's Cross. However, this was no true

This view shows the front section of the National Air and Space Museum's Me 262. The museum's aircraft was captured at Lechfeld, Germany, by a special team led by Col. Harold M. Watson, who directed Operation Lusty, the discovery and seizure of advanced German aircraft.

mutiny, just a desire to fly the most advanced aircraft available. Long after the war, Galland remarked that with just three hundred Me 262s at his disposal, he could have shot down two hundred Allied bombers per day, and that this would have halted the Allied bombing of Germany. Yet this was wishful retrospective thinking, which nevertheless fueled the everlasting discussion of the Me 262 program.

Furthermore, claims by Me 262 pilot Hans Guido Mutke that he and possibly other Me 262 pilots crossed the sound barrier created another myth. Messerschmitt company test flights showed that at speeds above Mach .83 the plane became increasingly unstable. Test pilot Gerd Lindner, who took the Me 262 to Mach .86 in strictly controlled research dives, experiencing violent buffeting and extremely alarming high-frequency pitching. Speeds beyond .86 would lead to a nose-down trim, which the pilot had no chance to counteract. An increasingly steeper dive would lead to even higher speeds and destruction of the airframe due to excessive negative G loads. These reports undermine any claims that Me 262s exceeded Mach 1, the speed of sound.

Despite its sleek appearance, the Me 262 was built in a primitive manner. The high number of unskilled laborers involved in aircraft manufacturing in decentralized production sites all over Germany required a serious simplification and standardization of the work process, which Messerschmitt had considered in his modular design philosophy. Additionally, the aircraft's components and subsections were not designed to close tolerances because discrepancies in the measurements of the various segments were common. In final assembly, a putty-like substance was generously applied to the resulting gaps, dents, and seams, then taped and painted over. During the restoration, these features were carefully recreated, since they provide an important glimpse into the aircraft's production history, including the employment of slave laborers.

The Me 262 was, like other so-called *Wunderwaffen* (wonder weapons), far less effective in service than Nazi-era propaganda suggested. It was a fighter built to hit and run in the final months of war. Realizing the aircraft's capabilities, Nazi propaganda went from hope to faith to delusion, yet the high expectations never materialized. The Me 262 came at a time when Germany was already moribund, facing a furious avalanche of Allied power. The final battles of World War II's European theater were not fought in the air but by infantry and tanks on the ground. Despite its breathtaking technology, the Me 262 was a marginal note in the history of World War II and one of the last technological achievements of the dying Third Reich.

But the Me 262 also provided a glimpse of the future. In the next big international conflict, the Korean War, the American F-86 Sabre and the Russian MiG-15 would face off. Both designs were based on careful studies of German jet technology and high-speed aircraft designs by the Allied victors. Finally, jet aircraft would dominate the skies over battlefields.

A view of the Me 262's cockpit after restoration. Due to lack of aluminum in war-torn Germany, the instrument board was made of plywood.

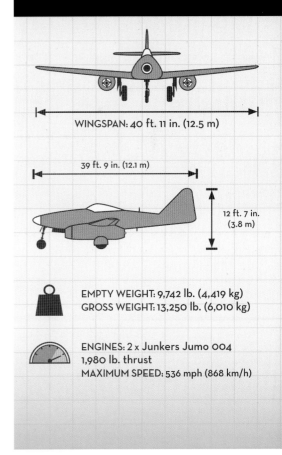

SPECIFICATIONS
MESSERSCHMITT ME 262A 1-A SCHWALBE

WINGSPAN: 40 ft. 11 in. (12.5 m)

39 ft. 9 in. (12.1 m)

12 ft. 7 in. (3.8 m)

EMPTY WEIGHT: 9,742 lb. (4,419 kg)
GROSS WEIGHT: 13,250 lb. (6,010 kg)

ENGINES: 2 x Junkers Jumo 004
1,980 lb. thrust
MAXIMUM SPEED: 536 mph (868 km/h)

BOEING B-29 SUPERFORTRESS
ENOLA GAY

CHAPTER 15

BY JEREMY R. KINNEY

World War II, the bloodiest conflict in human history, ended in the skies over Japan in August 1945. On August 6, the crew of a modified Boeing B-29 Superfortress named *Enola Gay* dropped an atomic bomb called "Little Boy" on the city of Hiroshima, killing tens of thousands of people instantly. Another atomic attack on Nagasaki followed three days later. On August 15, Japan announced its surrender.

The Superfortress was the most advanced propeller-driven airplane in the world—a technological marvel that represented the latest advances in American aeronautical engineering, bomber design, and strategic bombing doctrine. The *Enola Gay* facilitated a turning point in human history, ushering in the Atomic Age and the threat of nuclear war.

The B-29 originated from a February 1940 US Army Air Corps call for proposals for a very-long-range bomber with a pressurized cabin, tricycle landing gear, and the ability to carry a maximum bomb load of 2,000 pounds at a speed of 400 miles per hour to a target 2,500 miles away. Boeing, Consolidated Aircraft, Douglas, and Lockheed all responded with proposals. The air corps selected the Boeing entry and issued a contract for two prototypes to be delivered in September 1940.

In 2003, after a twenty-year restoration, *Enola Gay* was assembled for the first time since 1960 at the Steven F. Udvar-Hazy Center.

TIMELINE
BOEING B-29 SUPERFORTRESS
ENOLA GAY

1940
- February: US Army Air Corps issues request for proposal for very long-range bomber

1942
- September 21: First flight of XB-29

1943
- February: Engine fire causes fatal crash of second prototype, killing test pilot Eddie Allen

1944
- June: First B-29 raids conducted against Japanese targets in Thailand
- June 15: B-29 raid strikes Japanese mainland
- November 24: B-29s bomb Tokyo from new bases in Mariana Islands

1945
- March 9: Massive B-29 firebombing destroys much of Tokyo
- August 6: *Enola Gay* drops first atomic bomb, destroying Japanese city of Hiroshima
- August 9: B-29 *Bockscar* drops second atomic bomb, destroying Nagasaki

1946
- July: *Enola Gay* participates in Operation Crossroads nuclear tests in Pacific
- August 30: *Enola Gay* transferred to Smithsonian Institution

1953
- Dcember 2: Original commander Col. Paul Tibbets delivers *Enola Gay* to Andrews Air Force Base

1995
- Portions of the aircraft go on public display for the first time on the National Mall

2003
- *Enola Gay* goes on exhibit at Udavr-Hazy Center

Boeing's streamlined Model 345 design with its tubular fuselage represented the latest advances. The long, narrow, high-aspect-ratio wing equipped with large Fowler-type flaps permitted high airspeeds at high altitudes but maintained comfortable handling characteristics during the slower airspeeds necessary during takeoff and landing. For propulsion, the bomber relied on four high-output eighteen-cylinder, 2,200-horsepower Wright R-3350 radial engines and full-feathering, constant-speed propellers 16 feet 7 inches in diameter. The overall design could carry 16,000 pounds of bombs while cruising 235 miles per hour at altitudes up to 30,000 feet.

Boeing engineers also adopted several new innovative features to enhance operational capability. To protect the bomber against enemy fighters, they designed a weapons system that gunners operated by remote control. There were five turrets along the length of the fuselage: upper and lower forward turrets behind the cockpit, upper and lower rear turrets near the vertical tail, and a tail turret. Each featured two .50-caliber machine guns, with the forward upper turret adding two more in later versions and the tail turret incorporating an additional 20-millimeter cannon. Using computerized sights from their remote stations, each gunner could control multiple turrets to concentrate firepower as they tracked an incoming target.

Since the defensive system did not require open hatches or gunners to be located in turrets, the design also benefitted from another new innovation: three pressurized crew sections—a first for bomber aircraft—joined by a tunnel. The crews in the flight deck in the forward fuselage, the mid-fuselage compartment just aft of the wing, and the tail-gun position did not have to rely on oxygen masks or bulky clothing during noncombat

operations. Regulated cabin pressure and temperature would increase crew comfort on the long flights anticipated.

The Boeing design also included sophisticated radar equipment and avionics. An AN/APQ-13 or AN/APQ-7 Eagle radar system aided in navigation and accurate bombing even through cloud layers that completely obscured a target, but radio and radar operators had to operate up to twenty different types of radio and navigation equipment over the course of a single mission.

The prototype XB-29 took flight at Boeing Field, Seattle, on September 21, 1942. To build airframes, Boeing built two new plants at Renton, Washington, and Wichita, Kansas, while licensees Bell and Martin erected factories at Marietta, Georgia, and Omaha, Nebraska, respectively. Thousands of subcontractors produced smaller components and equipment. For the R-3350

radial engine, Wright Aeronautical constructed a new factory at Wood-Ridge, New Jersey, and dedicated all production at its Cincinnati facility to the engines. The Dodge-Chicago division of the Chrysler Corporation also produced R-3350s exclusively.

The R-3350 generated significant problems for the overall B-29 program due to its tendency to overheat and catch fire. One such fire led to the crash of the second prototype in February 1943, killing the entire crew, including Boeing chief test pilot Eddie Allen.

Continued changes to the overall design threatened the B-29's production schedule and operational introduction. Because production began before the US Army Air Forces finished testing, modification centers were established to effect last-minute design changes without halting overall production. The dramatic six-week period in March and April 1944 during which the USAAF rushed to get the first operational B-29s ready for service became known as the "Battle of Kansas." When production ended in 1946, Boeing, Bell, and Martin had built 2,766; 668; and 536 B-29s, respectively. Initially, the army air forces intended to deploy B-29s against Germany, but the production delays led to the decision to operate them in Asia and the Pacific. A significant component of the pre-invasion strategy was to weaken the Japanese ability to fight by attacking industrial and population centers. The key weapon in this strategy was the B-29.

The B-29 was operated by the Twentieth Air Force. Beginning in December 1943, Operation Matterhorn established bases in India and China to conduct operations in the region and against mainland Japan. The four bombardment groups of the Twentieth's XX Bomber Command arrived in April 1944 and flew their first combat mission, the longest of the war to date, in June against Bangkok, Thailand. The group suffered no losses, but twenty of the ninety-eight B-29s failed to reach the target, and bombing results were mediocre. The first raid against Japan occurred on June 15, 1944, but due to airfields in

Above: The forward cockpit of *Enola Gay* included (from left) the command pilot, bombardier, and pilot positions occupied by Paul Tibbets, Tom Ferebee, and Robert Lewis, respectively.

Opposite: *Enola Gay* on Tinian in the Mariana Islands.

Staff Sergeant Wyatt Duzenbury occupied the flight engineer station just behind the co-pilot's position, where he monitored the operation of the propellers and engines.

China that were vulnerable to Japanese attack and the lack of a direct supply line, all materials had to be flown over the Himalayas, severely limiting the effectiveness of the raids, which concluded in January 1945. The last B-29 mission flown from India occurred in March 1945.

The costly capture of Saipan, Tinian, and Guam in the Mariana Islands in August 1944 gave the army air forces bases several hundred miles closer to mainland Japan than those in India and China. The five bombardment wings of the Twentieth Air Force's XXI Bomber Command began to arrive in October 1944. On November 24, all four groups of the 73rd Bombardment Wing commenced operations against a Nakajima aircraft engine plant in Tokyo.

The operational environment over Japan deterred the USAAF from following its usual doctrine of daylight, high-altitude precision bombing. Weather in the northwestern Pacific made visual target acquisition nearly impossible. Additionally, the jet stream, the strong and steady current of winds found at high altitude, propelled B-29s quickly over their targets and overwhelmed bombing computers if pilots followed it or made the bombers motionless if they faced it head on. The primary architecture in Japan was light wood structures that high-explosive bombs rarely affected. Industries crucial to Japan's war effort were interspersed among civilian housing, and Gen. Haywood S. Hansell, commander of XXI Bomber Command, ordered his units to attack specific industrial and logistical targets and avoid population centers. This adherence to doctrine resulted in poor bombing results. General Henry H. "Hap" Arnold relieved Hansell in January 1945 due in part to his reluctance to adopt new methods reflecting the operational environment over Japan.

Hansell's replacement, Gen. Curtis E. LeMay, was a veteran bomber commander. Under him, the Twentieth aimed to destroy Japanese industry and kill or drive away its workers by burning the mostly wooden cities to the ground. LeMay switched the focus to low-level, nighttime firebombing raids. He ordered the removal of most of the B-29s' defensive armaments so they could carry more incendiary bombs and specified that their undersides be painted black. In March 1945, approximately three hundred B-29s attacked Tokyo, killing more than one hundred thousand people and destroying one-fourth of the city in a single twenty-four-hour period. Additional daylight raids and the aerial mining of shipping lanes left Japan isolated and in a military, economic, political, and social shambles by the end of July 1945.

As the American bombing campaign reached its crescendo, the United States was developing a new weapon. After a successful test explosion in July 1945 proved their capability, the US military had two new weapons to use against Japan and to affect a rapid end to the war: a uranium-235 gun-type fission weapon called "Little Boy," and a plutonium implosion-type called "Fat Man."

In June 1943 Project Silverplate began work to modify B-29s into atomic bombers. Silverplate B-29s had no armor plating and no upper or lower fuselage turrets, reducing the weight of the aircraft by 7,200 pounds. Reversible Curtiss Electric propellers provided backward thrust to slow the lumbering bomber on the runway if it had to land with the bomb. The forward bomb bay and forward wing spar required modification to accommodate a single bomb that would weigh in the area of 10,000 pounds. To that end, engineers adopted the British Type G single-point attachments and Type F releases used on the Avro Lancaster to carry the 12,000-pound Tallboy bomb. Overall, the changes enabled the bomber to carry an atomic bomb while cruising 260 miles per hour at 30,000 feet.

To deliver the weapon, the army air forces created the 509th Composite Group under the command of Col. Paul W. Tibbetts Jr., a battle-hardened B-17 veteran who chose fellow Eighth Air Force veterans Maj. Thomas Ferebee and navigator Capt. Theodore "Dutch" Van Kirk to join him in leading the group.

Production of the first fifteen Silverplate B-29s took place at the Glenn L. Martin factory in Omaha. Tibbetts personally selected one of them as his personal aircraft. On May 18, 1945, the army air forces received the B-29-45-MO with serial number 44-86292, and the 509th assigned it to crew B-9 commanded by Capt. Robert A. Lewis. They arrived at Wendover, Utah, on June 14 for training and departed for Tinian by way of Guam on June 27.

Crew B-9 and their B-29 arrived on Tinian on July 6 and flew further training and practice bomb missions, as well as two combat missions against Kobe and Nagoya using

Bottom left: This area just behind the rear crew department featured a camera bay and a black "putt putt" motor that provided power when starting the B-29.

Bottom right: Staff Sergeant George R. Caron occupied the tail-gun position, where he filmed the mushroom cloud forming over Hiroshima as *Enola Gay* turned back toward Tinian.

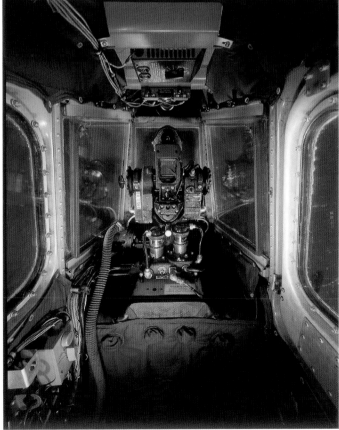

Maintenance officer Lt. Col. John Porter poses with the *Enola Gay* flight crew. Standing (from left): Porter, Capt. Van Kirk, Maj. Ferebee, Col. Tibbets, Capt. Lewis, and Lt. Jacob Beser. Front row (from left): Sgt. Joseph Stiborik, SSgt. George Caron, Pfc. Richard Nelson, Sgt. Robert Shumard, and SSgt. Wyatt Duzenbury.

"pumpkin bombs," Fat Man casings filled with 6,300 pounds of high explosives. Tibbets commanded a test mission that dropped a model of Little Boy near Tinian on July 31. Soon after, personnel added the "Circle R" of the 6th Bombardment Group on both sides of the vertical tail and the number 82 just behind the bombardier's window position to confuse enemy intelligence.

On Tinian, the 509th was under the operational control of the Twentieth Air Force headquartered in Washington, D.C. On August 2, the 509th issued Special Bombing Mission Number 13, designating Hiroshima as the target. The mission plan called for seven B-29s: three for weather reconnaissance over three potential targets one hour in advance, one to deliver the bomb, one to drop blast measurement packages simultaneously to record the effect of the bomb, one equipped with cameras for photographic documentation, and one assigned to the emergency field at Iwo Jima to serve as an alternate bomb carrier in the event of an emergency. On August 5, Tibbets ordered Allan L. Karl, an enlisted mapmaker

and off-duty sign painter, to add "Enola Gay" to the left side of the aircraft under the pilot's window in honor of his mother.

Tibbetts and his crew took off from Tinian at 2:45 a.m. on August 6, 1945. Van Kirk plotted the 1,500-mile route to Hiroshima, and they rendezvoused with the measurement and photographic B-29s over Iwo Jima. Weaponeer and mission commander Capt. Deak Parsons of the US Navy activated the bomb during the flight, and his assistant, 2nd Lt. Morris R. Jeppson, armed the safety plugs thirty minutes before reaching the target. Upon visual location of Hiroshima, Tom Ferebee aimed for the city center and *Enola Gay* dropped Little Boy from 31,000 feet at 9:15 a.m. Radar operator Jacob Beser tracked the bomb as it fell forty-three seconds to its predetermined detonation height approximately 2,000 feet over the city center. In an explosion equivalent to 16 kilotons of TNT, the bomb killed an estimated sixty thousand people instantly. Another sixty thousand died later from radiation sickness and related injuries. Little Boy destroyed 4.7 square miles of the city and left less than 20 percent of the city's buildings standing.

After Little Boy left the forward bomb bay, *Enola Gay* lurched upward and Tibbets initiated a high-angle evasive maneuver get as far away from Hiroshima as possible.

Enola Gay lands on the airstrip at Tinian after completing the mission to drop *Little Boy* on Hiroshima on August 6, 1945.

A bright flash overwhelmed the senses of the crew. The bomber traveled 11.5 miles before it experienced the shock waves from the blast; Staff Sgt. Robert Caron in the tail gun position took a photograph of the mushroom cloud over Hiroshima. Radar operator Sgt. Joe Stiborik recalled that the crew was speechless. Lewis wrote in his journal (and may have subconsciously said aloud over the radio intercom), "My God what have we done." *Enola Gay* landed at Tinian twelve hours later, at approximately 3:00 p.m. local time. Tibbets received the Distinguished Service Cross from Lt. Gen. Carl Spaatz upon arrival.

Three days later on August 9, Maj. Charles W. Sweeney and the crew of the B-29 *Bockscar* dropped Fat Man on Nagasaki, Japan. The bomb killed approximately thirty-five

The forward fuselage of *Enola Gay* is transported from the National Air and Space Museum on the National Mall in 2004.

thousand people instantly; another forty thousand died from sickness and injuries. Crew B-10, under the command of Capt. George Marquardt, flew *Enola Gay* as part of that mission, performing advance weather recon on the city of Kokura, which was the primary target.

On August 9, the Soviet Union declared war and invaded Japanese-held Manchuria. The end of the war in the Pacific was near. Facing invasion and the threat of continued incendiary and atomic bombings, Emperor Hirohito, in an unprecedented radio address to the nation, announced the Japanese government's intention to surrender to the Allies on August 14. A formal ceremony followed on September 2, 1945, aboard the battleship USS *Missouri* anchored in Tokyo Bay.

Controversy shrouds the history and memory of the atomic attacks on Hiroshima and Nagasaki and their effect on ending the war in the Pacific. Military and political leaders, as well as generations of veterans since, argued that using the atomic bombs against Japan shortened the war and prevented a large-scale Allied invasion of Japan that would have resulted in the loss of hundreds of thousands of lives on both sides. They saw the bitter resistance put forth by Japanese forces on islands like Iwo Jima and Okinawa and mass aerial suicide attacks by kamikaze units as indications of Japanese resolve to continue the war.

Critics of the decision to use atomic weapons recognized the terrifying power and responsibility created by what would be commonly referred to as "the bomb" in the postwar period. The bombings of Hiroshima and Nagasaki were a striking example of how technology, normally celebrated as a positive force before the war, could also bring utter destruction.

Lewis and crew B-9 ferried *Enola Gay* from Tinian to Roswell Army Air Field in New Mexico in November 1945, and Tibbets flew *Enola Gay* during Test Able of the Operation Crossroads nuclear weapons test program in the South Pacific on July 1, 1946. Upon its second return to the United States, *Enola Gay* became part of the large aircraft inventory stored at Davis-Monthan Army Air Field outside Tucson, Arizona. On August 30, 1946, the army air forces officially transferred *Enola Gay* to the Smithsonian Institution.

The newly created US Air Force moved the bomber, with Tibbets at the controls, to Orchard Place Air Field at Park Ridge, Illinois, outside Chicago in July 1949. The Smithsonian formally accepted the bomber and stored it with other aircraft in its possession there. When the institution lost the facility due to the impending construction of O'Hare Airport, the air force flew the bomber to Pyote Air Force Base, Texas, in January 1952. *Enola Gay* made its last flight, from Pyote to Andrews Air Force Base outside of Washington, D.C., on December 2, 1953. The bomber sat outside for almost seven years until Smithsonian staff began disassembly in August 1960 and transported the components to a storage facility in Suitland, Maryland, in July 1961.

National Air and Space Museum staff began restoration of *Enola Gay* in December 1984. It was the largest treatment project undertaken by the museum, requiring nearly two decades and approximately three hundred thousand hours to complete. The challenge included polishing virtually every square inch of the bare aluminum surface, refurbishing all four R-3350 engines, and sourcing components that went missing over the years.

The fully assembled artifact, the first time *Enola Gay* was a complete airplane since 1960, went on permanent display at the Steven F. Udvar-Hazy Center in December 2003.

SPECIFICATIONS
BOEING B-29 SUPERFORTRESS
ENOLA GAY

WINGSPAN: 141 ft. 3 in. (43.1 m)

99 ft. (30.2 m)

29 ft. 7 in. (9 m)

EMPTY WEIGHT: 69,000 lb. (31,400 kg)
GROSS WEIGHT: 135,000 lb. (61,363 kg)

ENGINES: 4 x Wright R-3350-41 Cyclone turbo-supercharged radial 18-cylinder, 2,200 hp each
MAXIMUM SPEED: 358 mph (576 km/h)

LITTLE GEE BEE

CHAPTER 16

BY RUSSELL E. LEE

Interest in providing aircraft for the nonprofessional amateur gained traction in the United States following two dramatic light-plane races held during the 1924 National Air Races. Future Boeing company test pilot Eddie Allen called the event "perhaps, the official inception of light [homebuilt] planes in America."

More than any other factor, however, dramatic reductions in the costs of building these aircraft made the sky accessible to vast new groups of people and drove the design of homebuilt aircraft. By the late 1920s, several professional designers were producing aircraft that amateurs could build from detailed plans or from manufactured components packaged as kits. Magazines such as *Modern Mechanix and Inventions, Science and Invention, Popular Mechanics,* and *Modern Mechanics* hawked the new low-cost flying technology by publishing race reports, colorful advertisements for plans and kits, and illustrated how-to articles. Although amateurs could build two-seat aircraft, they usually opted for single-seaters, while a growing number of amateurs used readily available engines and hardware to design and build their own unique aircraft.

Department of Commerce figures at the end of 1930 showed 2,464 unlicensed aircraft in the United States. Yet even as the amateurs seemed to prosper, the airline industry

Little Gee Bee is on display at the Smithsonian National Air and Space Museum's Udvar-Hazy Center.

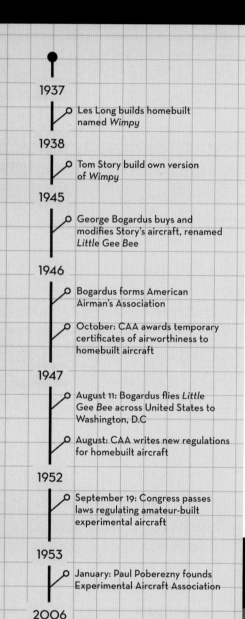

1937
- Les Long builds homebuilt named *Wimpy*

1938
- Tom Story build own version of *Wimpy*

1945
- George Bogardus buys and modifies Story's aircraft, renamed *Little Gee Bee*

1946
- Bogardus forms American Airman's Association
- October: CAA awards temporary certificates of airworthiness to homebuilt aircraft

1947
- August 11: Bogardus flies *Little Gee Bee* across United States to Washington, D.C
- August: CAA writes new regulations for homebuilt aircraft

1952
- September 19: Congress passes laws regulating amateur-built experimental aircraft

1953
- January: Paul Poberezny founds Experimental Aircraft Association

2006
- *Little Gee Bee* donated to National Air and Space Museum

was growing fitfully as potential customers refused to take flying seriously. Government regulation was needed to give the industry credibility and to stimulate and constructively channel its growth. In May 1926, President Calvin Coolidge signed the Air Commerce Act into law, the first national legislation enacted to regulate civil aviation. The law established federal jurisdiction over all aircraft operating between states but left unregulated homebuilt aircraft operating within states.

At first, states showed little interest in regulating the homebuilders within their borders. By March 1928 only ten states had enacted regulations governing amateur flying.

Small groups of enthusiasts continued to build and fly homebuilt aircraft throughout the 1930s, but the increasing threat of war led federal authorities to strengthen their control. Federal regulations issued in 1937 and 1938 made no provisions for certifying homebuilt aircraft built solely for recreation flying. Immediately after the Pearl Harbor attack, the federal government prohibited the flying of all private factory-produced sport aircraft and all homebuilt airplanes.

After the war ended four years later, most military pilots and flight crews discharged from service had no further interest in flight, and a highly anticipated boom in private aircraft manufacturing and sales went bust. The fortunes looked dire for scores of companies, so the federal government took steps to revive private flying. Oregon homebuilders sensed an opportunity to enlighten the Civil Aeronautics Agency (CAA) in Washington, D.C., about the progress made in homebuilt aircraft design and construction since the early 1930s.

Federal interest in supporting private flying came none too soon. The surge of support from Oregon homebuilders had barely halted an attempt in early 1945 by the state legislature to outlaw homebuilt aircraft. Spurred to action, resident George Bogardus led an exchange of letters between the homebuilders (himself included) and CAA officials in Washington describing the precarious state of homebuilding nationwide. Soon a draft proposal to define amateur flying in language adaptable to written legislation began to circulate. In early 1946, Bogardus announced that he had formed the American Airman's Association (AAA) and would begin publishing a magazine called *Popular Flying* in partnership with Joseph Yutz of Pottsville, Pennsylvania, who had published his own pamphlets about homebuilding called *Fly-Air*. Enthusiasts from as far as Australia joined the AAA, and the movement began to grow.

The CAA invited Bogardus to appear before the Civil Aeronautics Board in April 1946, and Bogardus made the road trip in his 1937 Chevrolet, accompanied by wife Lillian, a

AMERICAN AIRMEN'S ASSOCIATION
"We Represent the Experimental and Sportplane Builder"
BEAVERTON and TROUTDALE, OREGON

relative who helped pay for gas, and members of the AAA. They met CAA officials on April 15 and presented four proposals drafted with input from activist homebuilders in Oregon, Utah, Minnesota, and Pennsylvania. Bogardus agreed to submit a revised proposal to regulate homebuilt aircraft the following year. The AAA group also met with Ed Ryder, Al Voellmecke, and other staff who had built their own sport airplanes before the war. During the return trip to Oregon, the group detoured to discuss their mission with Ben Shupack, secretary of the Soaring Society of America, and with the editors of *Air Trails* magazine. Both organizations promised to write letters of support to the CAA. Motivated largely by Bogardus's visit, the CAA agreed in October to grant new temporary certificates of airworthiness to homebuilt aircraft that enthusiasts had built before the war. It was a modest but important first step toward recognizing the legitimacy of homebuilts by using federal legislation to create a permanent regulatory category for the aircraft.

As the homebuilt community began to stir amid these signs of support emanating from the CAA, Bogardus decided it was time to build his own airplane and fly it to Washington, D.C., "to show CAA officials—and also the media—that the new breed of competently made homebuilts were real and capable aircraft." Bogardus already had all the parts to assemble a small single-seat airplane designed by Tom Story; it was the end product, said Bogardus, of thirteen years of work by several other designers.

The project had begun in 1935 when Les Long published an article in *Popular Aviation* describing a series of experiments he conducted. Long rearranged the wing of a small airplane he had designed into different configurations using the same fuselage equipped with the same engine, then tested each configuration in flight: a wire-braced, low-wing configuration; a strut-braced, low-wing layout; a wire-braced, high-wing parasol; and a low-wing, strut-braced, gull-wing arrangement he called the "high-low." He found that the low-wing layout braced with wire weighed less than the other configurations and had superior handling qualities and performance. The bracing wires formed a strong, lightweight truss framework that passed over the top of the fuselage and extended through each wing and down to the landing gear struts. Using all the lessons learned from his experiments, Long completed an airplane called *Wimpy* two years later.

Above: George Bogardus revs the 65-horsepower Continental engine that propelled him and *Little Gee Bee* from Oregon to Washington, D.C., in 1947.

Opposite: In 1946, Borgardus rallied a group of local pilots who enthusiastically supported amateur homebuilders and flying for sport.

A small array of instruments was all that George Bogardus needed to fly *Little Gee Bee* at a cruising speed of 80 to 90 miles per hour. The word "experimental" had to appear prominently on the cockpit of all amateur-built aircraft.

Tom Story liked *Wimpy* so much that he designed and built a similar airplane in 1938. Story flew this aircraft for several years and then stored it until Bogardus bought it in 1945. Bogardus upgraded the 40-horsepower engine to 65 horsepower and added a compass, an extra fuel tank, and other accessories. He earned the new CAA temporary airworthiness certificate by flying the homebuilt, which he named *Little Gee Bee*, more than fifty hours on local and cross-country flights beginning early in 1947. After confirming the meeting with the CAA, Bogardus took off for Washington on August 11, 1947. Making his way across the country, he detoured to Hempstead, New York, on Long Island to meet fellow homebuilder Alexander "Jack" McRae.

McRae had graduated from the University of Michigan with an aeronautical engineering degree and worked at Stinson and Republic. By this time an employee of the Grumman Aircraft Engineering Corporation and a builder and pilot of homebuilt aircraft, he added considerable credibility to Bogardus's efforts to convince the CAA to enact legislation to support the homebuilt community. McRae was at the controls of his own Cessna 140 airplane as he flew in formation with Bogardus in the *Little Gee Bee* from Hempstead to Washington at the end of August. When CAA and CAB officials examined the *Little Gee Bee*, they were impressed by the safety and reliability of the design. They agreed to grant temporary airworthiness certificates good for six months not just for prewar designs, but for all homebuilt airplanes that qualified, and to begin writing legislation to create a permanent regulatory category based on the following criteria submitted by Bogardus, McRae, and other members of the AAA:

- Pilots will first fly their homebuilt craft for 50–100 hours under a temporary certificate to demonstrate their airworthiness.

- The pilot must have a private pilot's license.

- The CAA will issue a new and more permanent certificate allowing the homebuilt to fly in the new category, once the demonstration of airworthiness concludes without encountering significant problems.

- The pilot/builder will placard the cockpit of two-place homebuilt aircraft to inform the passenger they are about to fly in a homebuilt airplane. [The CAA eventually chose the term "experimental" and instructed the homebuilder to apply the placard to both sides of the cockpit.]

- When someone buys a homebuilt aircraft, they must again fly for 50–100 hours to demonstrate airworthiness.

- Pilots cannot fly homebuilt airplanes over densely populated areas.

As the decade drew to a close, the CAA continued to favor the homebuilt movement. Bogardus flew *Little Gee Bee* to Washington again in 1949 and 1951. He contributed photographs and information about past and current homebuilt airplanes to an article that appeared in the popular *Air Trails* magazine in August 1950, casting national attention on homebuilt aircraft. Charles E. Planck, chief of the CAA Information Division, wrote in the same magazine: "CAA has no written specifications for homebuilt aircraft . . . [but it] is possible that we will have more definite standards for special-purpose airplanes within a couple of months. . . . When and if that regulation becomes effective, we would be able to establish and publish standards for the amateur-built plane in more specific terms than we can now."

A proposal to revise the CAA regulations to include a more detailed definition of amateur-built airplanes appeared in the Federal Register on November 17, 1951. Finally, on September 19, 1952, Congress enacted the law creating a permanent regulatory category for amateur-built experimental aircraft. Four months later, Paul Poberezny and a small group of enthusiasts founded the Experimental Aircraft Association (EAA), named for the word that was required to mark the cockpit of every homebuilt aircraft. In 2015, the EAA claimed 180,000 members worldwide. Today, people who build and fly homebuilts include active and retired military and commercial pilots, professional businessmen and pilots of corporate business aircraft, astronauts, scientists, professors, members of the clergy, ranchers, mothers, daughters, fathers, and sons. All share several traits: an independent spirit, confidence in their ability to read and understand an instruction manual, the eye-hand coordination needed to wield the tools required to assemble an aircraft, and the tenacity and focus required to complete a fairly complex years-long project.

Available to this diverse group of pilot-builders are homebuilt airplanes of many types and configurations. Most seat just the pilot, but others can carry five or six people in airframes made of wood, metal, and composite materials that combine synthetic foam covered with fiberglass or carbon fiber. The main wings are not always in front, sometimes being installed at the rear behind a small wing, or canard. There are monoplane, biplane, and triplane homebuilts, some with retractable landing gear and others with fixed gear. Reciprocating engines fueled with gasoline power most homebuilts, but a few use jet engines. Some homebuilt airplanes are designed to fly fast and high, others low and slow, but regardless of their performance capabilities, all trace their roots back to the *Little Gee Bee*.

"*Little Gee Bee* is probably the most significant single homebuilt ever produced in the United States," Peter Bowers later wrote in *Guide to Homebuilts, 9th Edition*. "Its demonstrated reliability on a round-trip from Oregon to Washington, D. C., where George [Bogardus] petitioned the government for recognition of homebuilts, resulted in the establishment of the amateur-built category that we have today."

George Bogardus died in 1997 and left his estate, including the dismantled *Little Gee Bee*, to Chapter 105 of the EAA in Portland, Oregon. Chapter members, led by homebuilt airplane designer Richard "Dick" VanGrunsven, began restoring the *Little Gee Bee* in January 2005. After completing the project, except for paint and markings, the group generously donated the airplane to the Smithsonian National Air and Space Museum in November 2006. Museum treatment specialists Jeff Mercer and Bob Wyrock painted the various markings that Bogardus had applied to the original fabric covering the *Little Gee Bee* before his 1947 flight to Washington, D.C. Today the airplane is displayed in the Boeing Aviation Hangar at the Steven F. Udvar-Hazy Center.

SPECIFICATIONS
LITTLE GEE BEE

WINGSPAN: 29 ft. 8 in. (9 m)

19 ft. (5.7 m)

5 ft. (1.5 m)

EMPTY WEIGHT: 485 lb. (220 kg)
GROSS WEIGHT: 743 lb. (337 kg)

ENGINES: 1x Continental A-40 air-cooled 4-cylinder, 65 hp
MAXIMUM SPEED: 121 mph (195 km/h)

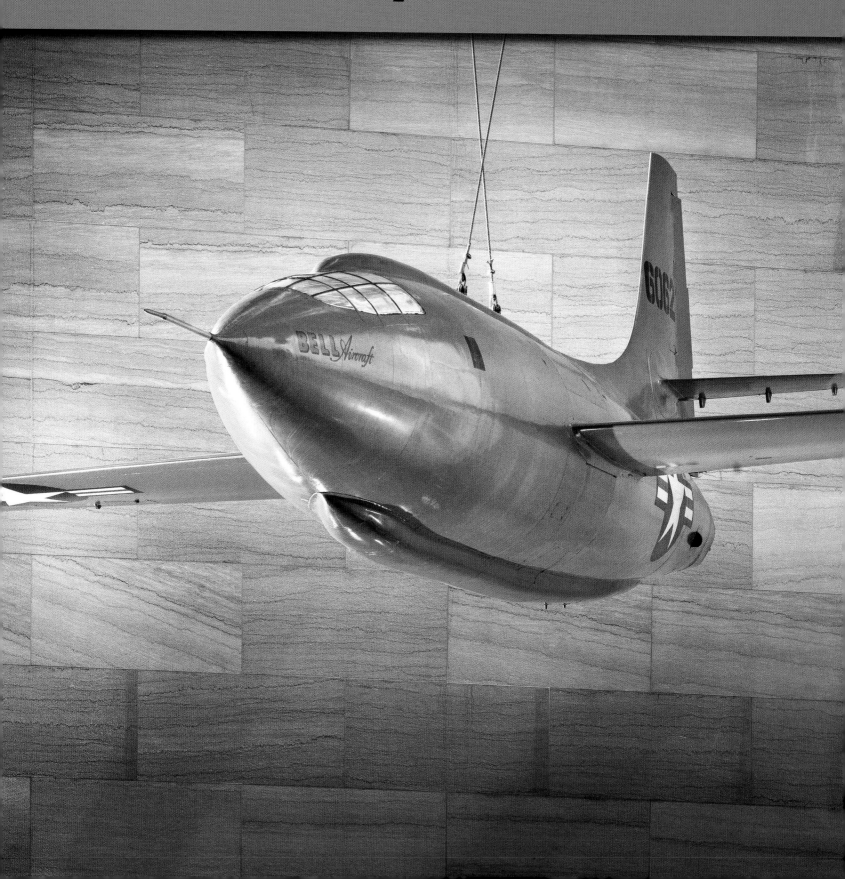

BELL XS-1/X-1

CHAPTER 17

BY DOMINICK A. PISANO

For aeronautical experts in the mid-twentieth century, to fly faster than the speed of sound was thought to be a momentous technological achievement. The National Air and Space Museum's Bell X-1 (or XS-1), was a combined development of Bell Aircraft, the US Army Air Forces/US Air Force, and the National Advisory Committee for Aeronautics (NACA). The X-1 was instrumental in gathering pioneering high-speed aeronautical research data, and on October 14, 1947, succeeded in breaking the sound barrier.

When aeronautical engineers began designing aircraft that approached the speed of sound, they encountered what became known as compressibility—aerodynamic effects that seemed to impede faster speeds. The speed of sound varies according to altitude and temperature; because of this variation, the speed of sound at any altitude is referred to as Mach 1, so named for Austrian physicist Ernst Mach, who first identified the phenomenon when studying the effects of shockwaves on ballistics. Between 40,000 and 60,000 feet at a temperature of -56.6 degrees Celsius (-69.88 degrees Fahrenheit), the speed of sound is a constant 659.8 miles per hour.

In aeronautical terms, these shockwaves in the transonic region (a range of speeds below, at, and beyond the speed of sound) were first discovered in the fall of 1918 by two

The Smithsonian National Air and Space Museum's Bell X-1, *Glamorous Glennis*, as displayed in the Milestones of Flight Gallery, is the first aircraft to fly faster than the speed of sound.

1944

○ Transonic research aircraft proposed by Ezra Kotcher of US Army Air Forces' Wright Field Design Branch

1945

○ November: US Army Air Forces approve production of Bell XS-1

1946

○ January 25: First glide test by pilot Jack Woolams

1947

○ April 11: Chalmers "Slick" Goodlin makes first powered flight in XS-1

○ October 14: Captain Chuck Yeager breaks sound barrier in X-1

1948

○ December 17: X-1 team awarded Collier Trophy

1950

○ Yeager pilots X-1 on last flight for movie *Jet Pilot*

○ August 26: X-1 transferred to Smithsonian Institution

engineers, Frank W. Caldwell and Elisha N. Fales, at the Army Air Service Engineering Division while conducting wind-tunnel tests on propellers. Caldwell and Fales determined that at high speeds, compressibility could be observed to affect the airflow around the propellers being tested.

Technology historian James R. Hansen writes that the conception of the "sound barrier" is owed to British aerodynamicist W. F. Hilton, who, in trying to explain the features of a subsonic wind tunnel, calculated it would take more than 30,000 horsepower to exceed the speed of sound. For aircraft, the speed of sound, Hilton feared, would be a "barrier against future progress." Many aerodynamicists believed that the "sound barrier" made going faster than sound improbable if not impossible.

One theorist, however—John Stack, a young aeronautical engineer at the NACA Langley Memorial Aeronautical Laboratory in Hampton, Virginia—believed that the effects of compressibility on aircraft could be overcome. Stack's paper entitled "Effects of Compressibility on High-Speed Flight" was published in January 1934 in *The Journal of the Aeronautical Sciences*, and his subsequent work was seminal in the development of the Bell X-1.

The inspiration for the Bell X-1 was a design proposal by the Wright Field Design Branch of the US Army Air Forces Aircraft Laboratory, which in turn was based on a study done in early 1944 by Ezra Kotcher titled "Mach 0.999." Kotcher, who held a position in the Engineering Division at Wright Field, had graduated from the University of California in 1928 and gone to work as a senior instructor at the Air Corps Engineering School at Wright Field. In 1939, Kotcher had prepared a brief report for Maj. Gen. Henry H. "Hap" Arnold—then chief of the USAAF's predecessor, the US Army Air Corps—that suggested investigations into

transonic research would require an extensive flight research program using rocket- or gas-turbine-propelled aircraft.

The design proposal concluded that a rocket-propelled aircraft was the most likely type to achieve transonic speeds. This hypothetical aircraft had a fuselage that was circular in cross-section, with relatively straight wings mounted in the middle of the fuselage and a conventional tail section. The cockpit canopy would be faired and powered by a liquid-fueled rocket engine generating 6,000 pounds of thrust. The proposal's similarity to the eventual Bell XS-1 is remarkable.

The design specifications for the high-performance aircraft were offered to Republic Aviation, which had built the P-47, and North American Aviation, which had built the P-51 Mustang, because both of these advanced fighters had encountered compressibility. At the time, neither company was capable of allotting engineering staff to design the aircraft.

A third company, the Bell Aircraft Corporation of Buffalo, New York, had produced the Bell XP-59 and P-59A Airacomet, the United States' first jet-powered aircraft, which had first flown October 1942 from Muroc Army Air Field (now Edwards Air Force Base) in the California desert, under top-secret conditions. In November 1944, Bell's chief design engineer, Robert J. Woods, was invited to Wright Field to discuss new fighter projects. While there, he stopped by Ezra Kotcher's office. Kotcher asked him if Bell might be interested in designing and building a high-performance, rocket-powered aircraft that would be purely for research, able to reach 800 miles per hour, and capable for flying at 35,000 feet for two minutes. When Woods returned to Buffalo, he began to assemble the team that would design and build what became known as the Bell XS-1/X-1: Paul Emmons, Benson Hamlin, Roy Sandstrom, and Stanley Smith. By the end of 1944, the aircraft had begun to take shape.

The X-1 was revolutionary in every way. Because there was little data in regard to transonic flight, the engineers began to tour various research facilities in search of information on how the fuselage should look. With engineers at the Ballistics Laboratory at Wright Field, they determined that the aerodynamic characteristics of high-powered bullets were a good place to start because these projectiles were known to travel at the speed of sound. Thus, the X-1's fuselage took on a bullet-shaped appearance. The fuselage was of conventional stressed-skin structure, but designed to withstand loads of +/-18 g (the force of gravity on Earth is 1 g).

Next, the Bell engineers took on the wings and tail section. There was disagreement about the shape of the wing (straight or swept) and about the thickness-to-chord ratio (i.e., the relationship between the maximum thickness or depth of an airfoil section and its chord—the distance from the leading edge to the trailing edge—expressed as a percentage of the chord length) of the wing and tail section. The engineers decided on a straight wing with a 10 percent thickness-to-chord ratio and a tail section with an 8 percent thickness-to-chord ratio.

The X-1 mockup was inspected by representatives from the NACA and the US Army Air Forces on November 10, 1945, and approved for production. On December 12, the X-1 was unofficially rolled out of the Bell factory in Wheatfield, New York; the official rollout took place on December 27. It was painted bright orange, so it could be easily identified, and assigned serial number 46-062.

The aircraft's rocket-powered engine (6000C-4/XLR-11) was still being furiously developed by Reaction Motors Inc. (RMI) of Pompton Plains, New Jersey, one of the

Above: Ezra Kotcher in the late 1950s. Kotcher's 1944 study entitled "Mach 0.999" helped inspire the X-1.

Opposite: NACA test pilot Chalmers "Slick" Goodlin is flanked by Reaction Motors founder Lowell Lawrence Jr. (left) and Bell Aircraft test pilot Dick Frost.

first US companies to build liquid-fueled rocket engines. The 6000C-4/XLR-11 consisted of four combustion chambers that used a combination of ethyl alcohol and liquid oxygen as propellants. It was capable of generating 6,000 pounds maximum thrust (1,500 pounds in each chamber), achieved by firing each chamber on and off individually.

Even though the engine aircraft was still in development, Bell engineers believed that valuable test data could be obtained in unpowered drop tests from the belly of its "mothership," a Boeing B-29 (the only American aircraft that could carry the X-1 to altitude) specially modified to accommodate the aircraft and conduct glide tests and subsequent powered flights. On January 19, 1946, the X-1 was transported by a B-29 to Pinecastle Army Air Field near Orlando, Florida, for initial glide tests. On January 25, the X-1, piloted by Bell Aircraft test pilot Jack Woolams, was carried to an altitude of 27,000 feet and released. The first

Above: Test pilot Chuck Yeager (right) and Paul E. Garber, head curator of the National Air and Space Museum, inspect the aircraft's 6000C-4 engine exhaust chambers.

Below: Robert "Bob" Hoover sits in the cockpit of a Lockheed P-38 Lightning at Wright Field, Ohio, in 1946. Hoover was an integral part of the X-1 flight test program.

unpowered flight went smoothly. In his report, Woolams wrote, "Of all the airplanes the writer has flown, only the [Bell] XP-77 and the Heinkel 162 compare with the XS-1 for maneuverability, control relationship, response to control movements, and lightness of control forces."

After ten glide flights were completed at Pinecastle, the X-1 was taken back to Bell Aircraft for modifications to its wings and tail section. The original configuration of 10 percent thickness-to-chord ratio on the wing was replaced by an 8 percent ratio, and the tail section's original 8 percent thickness-to-chord ratio was changed to 6 percent. The aircraft was then taken to for its first powered test flights to Muroc Army Air Field, chosen because of its remote location in California's Mojave Desert and because a large dry lake bed located there (now Rogers Dry Lake) had a surface area of 65 square miles—large

enough and secluded enough so that there would be plenty of space for emergency landings and no worries about flying over populated areas.

Chalmers "Slick" Goodlin replaced Woolams as Bell's chief test pilot. (Woolams was killed over Lake Ontario while test-flying a modified Bell P-39 configured for racing.) Subsequent flight tests were made through June 1947 to test the aircraft's handling qualities, buffet-boundary limits, and airspeed calibration. On April 11, 1947, Goodlin made the first powered flight in the X-1, and on June 5, 1947, he flew the X-1 in a demonstration flight for the Aviation Writers Association. In all, Goodlin flew nine test flights in 46-062 and eighteen in the second X-1, serial number 46-063.

That same month, USAAF and NACA representatives met at Wright Field to discuss the future of the X-1 program. The army air forces were not satisfied that sufficient progress was being made toward the goal of supersonic flight. Thus, the two organizations decided to go their separate ways, with the USAAF investigating the transonic-supersonic potential of the X-1 using 46-062 with the goal of breaking the sound barrier and NACA exploring the transonic control and stability using 46-063.

One of the men named to head up the USAAF X-1 test program was Col. Albert Boyd, chief of the Air Materiel Command's Flight Test Division and a veteran test pilot who in 1947 had set an absolute speed record of 623.738 miles per hour in a Lockheed P-80R. Army air force test pilots were asked to volunteer to fly the X-1 in the USAAF's attempt to break the sound barrier. Boyd selected one of the volunteers, Capt. Charles E. "Chuck" Yeager, a fighter pilot combat veteran who had distinguished himself during World War II with eleven and a half aerial victories, five of them in one day. For Yeager's alternate, Boyd chose Lt. Robert A. "Bob" Hoover, another seasoned fighter pilot. Rounding out the team as flight-test project engineer was Capt. Jack L. Ridley, a military pilot and brilliant aeronautical engineer with a master's degree from the California Institute of Technology and extensive experience in combat aircraft development.

Yeager flew the Air Force's X-1 (which he had named *Glamorous Glennis* after his wife) eight times for pilot familiarization, stability and control, and to check elevator and stabilizer buffet before his historic October 14 supersonic flight. On his October 5 test flight, he encountered shockwave buffeting at .86 Mach. Subsequently, the X-1's right wing dropped, and the aircraft began to vibrate. When Yeager increased his speed to .88 Mach, the X-1's ailerons (wing surfaces that control the aircraft's rolling motion) began to pulsate. Yeager had difficulty keeping the aircraft flying level.

On his next flight, once he gained altitude, Yeager fired each of the X-1's four rocket chambers. Once again, when he reached .86 and .88 Mach the aircraft was subjected to the same buffeting effect as on the previous flight. This time, however, as the aircraft

Above: X-1 flight test engineer Jack Ridley strolls in front of the the the X-1's Boeing B-29 mothership.

Below: Colonel Albert Boyd, chief of the Air Materiel Command's Flight Test Division, headed up the X-1 test program for the USAAF.

reached .94 Mach, he found that the X-1's elevators (tail surfaces that control the aircraft's nose-up and nose-down attitude) lost effectiveness. A disconsolate Yeager landed the X-1, believing that the aircraft could approach, but never exceed, the speed of sound.

The source of the problem, however, was soon discovered. The changes in the X-1's wing and tail thickness-to-chord ratios (to 8 and 6 percent, respectively) had caused the thinner horizontal stabilizer to encounter shockwave buffeting *after* the wing did, instead of simultaneously. The solution was to experiment with a horizontal tail that could be adjusted in increments by the pilot in flight.

On October 14, 1947, as Yeager made preparations for his historic flight, he was impaired by intense pain as a result of two broken ribs suffered on a horseback ride two nights before. Jack Ridley came to the rescue by fashioning the end of a broom handle into a lever that Yeager could use to secure the cockpit hatch of the X-1, which was in the belly of the B-29 mothership. The X-1 was now ready to fly. With Maj. Robert L. Cardenas at the controls of the B-29 and Lieutenant Hoover flying the Lockheed F-80 chase plane, the X-1 was released at approximately 20,000 feet. This was not done at the correct 260 miles per hour, however, and Yeager had to put the aircraft into a controlled dive to gain airspeed.

Once at speed, he fired the X-1's four rocket engines and climbed to 36,000 feet, hitting .88 Mach. He then shut down two of the engines to conserve fuel and climbed to 42,000 feet, where he hit Mach .92. He then refired the two inert engines. At that altitude the Mach meter registered .956 Mach and then Mach 1.06—700 miles per hour. The transition to supersonic speed was remarkably uneventful. The sound barrier had been broken.

At first the US Air Force, which the previous September had become a separate branch of the armed forces, did not want to publicize that Yeager and the X-1 team had

Opposite left: An interior view of the X-1 shows the instrument panel and control yoke.

Opposite right: Captain Chuck Yeager with the Bell X-1.

Below: From left, Capt. Jack Ridley, Capt. Charles "Chuck" Yeager, and Bell Aircraft test pilot Richard "Dick" Frost chat alongside the Bell X-1 during the aircraft's flight test program.

broken the sound barrier, but *Aviation Week* reported the event in its December 22, 1947, issue. Then, on March 26, 1948, with Yeager again at the controls, 46-062 attained a speed of Mach 1.45 (957 miles per hour) at an altitude of 71,900 feet, the highest velocity and altitude reached by a manned airplane up to that time. In June 1948, the USAF finally relented, admitting to these feats. Lawrence D. Bell, head of Bell Aircraft, John Stack of the NACA, and Yeager received the Collier Trophy (the pinnacle of aviation awards) in an official White House ceremony on December 17, 1948. Yeager, who also was awarded the Mackay Trophy for "most meritorious flight of the year" in late 1948, appeared on the April 18, 1949, cover of *Time* magazine.

Although Yeager lapsed into relative obscurity after breaking the sound barrier, he gained new celebrity with the publication of Tom Wolfe's book *The Right Stuff* in 1979 and the release of the film adaptation in 1983.

Although X-1 46-062 flew for a relatively short time, the X-1 program continued well into the 1950s, gathering essential information for the transonic flight regime. The air force and navy developed a number of jet-powered aircraft that could fly well beyond Mach 1 (the idea being that supersonic military aircraft could quickly attack incoming enemy aircraft), but as time went on and the nature of aerial combat changed, flying faster than the speed of sound became less important than other factors. In addition, the amount of fuel expended at supersonic-plus cruising speeds was prohibitively expensive.

Bell X-1 46-062 made its last flight on May 12, 1950, to record camera footage for *Jet Pilot*, an RKO film production supervised by Howard Hughes. The aircraft was officially accepted into the Smithsonian Institution's collections on August 26, 1950, in a ceremony presided over by Alexander Wetmore, secretary of the institution, and Hoyt S. Vandenberg, the chief of staff of the USAF.

SPECIFICATIONS
BELL XS-1/X-1

WINGSPAN: 28 ft. (8.5 m)

30 ft. 11 in. (9.4 m)

10 ft. (3.3 m)

EMPTY WEIGHT: 7,000 lb. (3,175 kg)
GROSS WEIGHT: 12,225 lb. (5,545 kg)

ENGINES: 1 x Reaction Motors XLR-11-RM3 4-chambered liquid-propellant rocket, 6,000 lbf (26.7 kN) total
MAXIMUM SPEED: 957 mph (1,540 km/h)

NORTH AMERICAN
F-86A SABRE

CHAPTER 18

BY THOMAS J. PAONE

Few aircraft have better defined a conflict more than the F-86 Sabre defined the Korean War. In the first meeting of swept-wing, jet-powered aircraft, the F-86 dominated the skies over the Korean peninsula and produced the incredible victory ratio of ten to one. Although it was radically redesigned during the early days of its development, it proved the value of swept-wing fighter aircraft and changed the art of aerial warfare.

The F-86 was born out of the boom of American jet-powered aircraft development in the last years of World War II, including the Bell XP-59A and Lockheed XP-80. The aircraft started out as North American Aviation (NAA) project NA-134 for the US Navy and was NAA's first attempt at developing jet aircraft. Work on the project began in the closing months of 1944, and engineers at NAA produced a conservative design intended to use the General Electric J-35 jet engine and a thin, straight wing.

While the NAA team was working on the proposal issued by the US Navy, the US Army Air Forces put forth a design request for a multi-use fighter capable of traveling at 600 miles per hour. On November 22, 1944, NAA proposed to the US Army Air Forces a modified design of the aircraft it was

Wind-tunnel tests for the F-86, such as this one at the 40-by-80-foot facility at the National Advisory Committee for Aeronautics (NACA) Ames Aeronautical Laboratory, helped prove and further develop the benefits of the swept-wing design.

1944

○ November 22: North American Aviation offers modified Project NA-134 for US Navy to Army Air Forces

1945

○ January 1: US Navy orders three XFJ-1s while US Army orders three similar XF-86s

○ November 1: Swept-wing version of XP-86 accepted by Army Air Forces

1947

○ October 1: First flight of XP-86

1948

○ May 28: First flight of J47-powered production XP-86

○ June: XP-86 redesignated XF-86

○ September 15: Speed record set by F-86A

1950

○ December 17: Lieutenant Colonel Bruce Hinton downs MiG-15 in first aerial combat between swept-wing fighters

1953

○ May 18: Jackie Cochrane, in Canadair-built F-86, becomes first woman to fly supersonic

1962

○ January 25: F-86A transferred to Smithsonian

developing for the navy. On January 1, 1945, the navy ordered three prototypes of the proposed aircraft, called the XFJ-1, while the army air forces ordered three prototypes that they designated the XP-86. In May 1945, the navy ordered production-model aircraft designated the FJ-1 Fury, which became the first operational jet-powered aircraft to enter service aboard aircraft carriers. The XP-86, however, would receive a major redesign before becoming one of the greatest jet fighters in US Air Force history.

After the prototype XP-86s were built, testing showed that the straight-wing aircraft could not meet the minimum speed requirements listed by the US Army Air Forces. Early wind-tunnel testing showed that the airframe design affected the maximum speed of the aircraft, requiring modifications if the aircraft was to compete with other designs submitted for consideration. At the time the XP-86 was being tested, captured German reports were arriving back in the United States showcasing German aeronautical testing of swept-wing aircraft. German wind-tunnel tests in 1942 had provided evidence of performance improvement to the first operational German jet fighter, the Me 262, when the wing was swept back, although it also introduced issues with stability.

To overcome the instability produced by the swept wing at lower speeds, German engineers had tested placing slats on the leading edges of the wings. These slats would be kept closed when the aircraft moved at high speeds and opened at slower speeds. Tests showed that this modification could dramatically improve the performance of an aircraft at low speeds while allowing it to still benefit from the swept-wing configuration at high speeds. In mid-1945, engineers at NAA, including Ed Horkey and Larry P. Greene, proposed pushing to reconfigure the design of the XP-86 to include swept wings. In August 1945, NAA received a grant from the US Army Air Forces to study the possibilities of a swept-wing design under design study RD-1369. In September 1945, engineers tested a scale model of the XP-86 with swept wings and wing slats in a wind tunnel, showing that performance, including speed, was greatly improved with the new design. On November 1, the US Army Air Forces accepted the design reconfiguration of the XP-86, and the first US swept-wing fighter was born.

The redesigned XP-86 continued to undergo design developments into 1946. The wing redesign caused delays to the program, but the US Army Air Forces was willing to accept the delays in exchange for the promise of improved performance. As a result, the leading-

edge slats were developed to be completely automatic. When the aircraft was traveling at high speeds, the slats closed without pilot intervention, minimizing their drag on the aircraft. At lower speeds, the slats opened and provided more lift. By June 1946, the final aircraft designs were approved, and construction began on the production prototype.

The first production prototype was completed in August 1947 and featured a J35-C-3 engine built by Chevrolet and capable of producing 4,000 pounds of thrust. The aircraft was transported to Muroc Air Field in southern California, where the US Army Air Forces' first military jet, the XP-59A, was tested during World War II. The first flight of the XP-86 was completed on October 1, 1947, by NAA test pilot George Welch, who had gained fame during World War II for being one of the first USAAF pilots to take off and shoot down a Japanese aircraft on December 7, 1941, after the attack on Pearl Harbor. After the first flight, Welch stated, "The plane's so clean you never have trouble. Reduce drag to a minimum and you don't have to worry about effects of compressibility shock waves. Spin recovery is easy when pressure on the elevators is released. It seems at first as if the nose is pointed too high and you might stall. You soon get used to it though."

The prototype was also noted by Welch as being underpowered, but the more powerful GE J47 engine was proposed for the full-production P-86 aircraft, solving the issue. Major Ken Chilstrom of the newly independent US Air Force flew the next flights of the aircraft for the second phase of testing, and improvements were made for the remainder of the testing period. On May 20, 1948, the first production models, powered by the J47, were officially accepted by the US Air Force, and in June 1948, the "P" (for "pursuit") designation previously given to the aircraft was changed to "F" (for "fighter").

Once the production model F-86 was available, the US Air Force wanted to showcase the capabilities of the new aircraft with an attempt to break the air-speed record. In 1947, the US Navy had broken the airspeed record with the specialized Douglas D-558-1 Skystreak. At the 1948 National Air Races in Cleveland, Ohio, Maj. Robert L. Johnston flew a fully loaded F-86A before thousands of spectators. Due to issues with the timing equipment, however, none of the flights could be recorded, preventing official breaking of the record. On September 15, 1948, Major Johnson made another attempt at the record over Muroc Air Base and officially set a new record of 670.981 miles per hour.

The F-86 also helped make history in the civilian world. On May 18, 1953, using a borrowed Canadian-built Canadair F-86 and operating under the tutelage of former X-1 flier Chuck Yeager, test pilot Jackie Cochran became the first woman to fly faster than the speed of sound, in addition to setting several speed records for flights over closed courses. In her autobiography, Cochran later wrote, "Believe me, breaking the sound barrier and being the first woman to do it was the greatest thrill of my life. The bulletlike shocks left the black and blue imprint of the shoulder straps on my body." Although the record-breaking attempts helped introduce the air force's new aircraft to the country, the first true test for the jet-fighter would come with the Korean War.

On June 25, 1950, North Korea launched a massive invasion of South Korea in an attempt to unify the country under a communist government. The situation quickly

Above: A North American Aviation XP-86 Sabre jet in flight over the Mojave Desert, California, in 1949. The swept-wing configuration greatly enhanced the speed of the aircraft.

Opposite: Armed with six .50-caliber machine guns, the F-86 Sabre proved one of the best fighter jets of the Korean War. This Sabre is part of the collection of the Smithsonian National Air and Space Museum.

deteriorated, and by the third day of the war, defenses around Seoul, the capital of South Korea, had nearly fallen and the evacuation of the country was underway. Although the United States had hoped to prevent the use of American ground forces on the peninsula, troops were sent and soon found themselves entangled in a conflict they were unprepared to fight. As the early days of the war brought one of the largest retreats in US military history, Gen. Douglas MacArthur turned to the air force to help turn the tide of the North Korean advance. On June 29, 1950, President Truman gave permission for the use of air operations north of the thirty-eighth parallel "to provide the fullest possible support to South Korean forces."

Once the US Air Force was committed to the Korean War, a combination of old and new aircraft entered the fight from forces stationed in the Pacific known as the Far East Air Force (FEAF). These included piston-engine-powered fighters and bombers from World War II as well as jet-powered fighters such as the P-80 Shooting Star. This initial commitment greatly helped United Nations ground forces push the North Koreans back to the Yalu River, the border between Korea and China. The situation changed drastically in October and November 1950, however, when Chinese forces entered the conflict, bringing not only great numbers of troops but a new fighter plane that would change the aerial conflict: the MiG-15.

The Mikoyan-Gurevich MiG-15 was designed in the Soviet Union in 1946 as a high-altitude, jet-powered interceptor. It was first flown on December 30, 1947, and featured swept wings similar to those of the F-86. After the success of the first test flights, the MiG-15 was quickly mass-produced and armed with one 37-millimeter cannon and two 23-millimeter cannons powerful enough to shoot down large bombers. After the invasion by Chinese forces, MiG-15 fighters soon appeared over the skies of Korea. On November 1, 1950, six of them flew over the Yalu River and challenged a flight of World War II–vintage P-51 Mustangs. The American craft were able to escape to their base, but the encounter showcased the new threat to American forces.

The MiG-15 fighters proved to be far superior to the P-51 fighters and B-29 bombers as well as 100 miles per hour faster than the F-80s. On November 8, 1950, MiG-15 aircraft engaged F-80C aircraft in the first all-jet battle in history. Though the F-80 was inferior, the combat experience of American pilots allowed Lt. Russell J. Brown to score the first aerial victory over a jet fighter in the Korean War. As more and more MiG fighters became operational, bombing missions by the US Air Force became more difficult. Older and slower propeller-driven B-29s lost the luxury of flying over a target multiple times before releasing bombs. The MiGs forced these bomber crews to act quickly, resulting in less accuracy and more sorties.

As UN ground forces retreated under the weight of the Chinese attacks, the FEAF fought to maintain the air superiority they had enjoyed for most of the war. To counter the new threat posed by the MiG-15, the US Air Force in November 1950 ordered the 4th

Opposite Top: Ground crew prepare a line of F-86 Sabres for combat in June of 1951.

Opposite bottom: After World War II, jet fighters captured the public's imagination. The F-86 co-starred with John Wayne and Janet Leigh in the 1957 Howard Hughes–produced film, *Jet Pilot*.

Left: The Mikoyan-Gurevich 15, or MiG-15, was the primary adversary of American F-86 Sabres over the skies of Korea. This MiG-15, formerly of the Chinese military, resides with the National Air and Space Museum's collection.

Fighter-Interceptor Wing, equipped with F-86A aircraft, to transfer to combat operations on the Korean peninsula. The wing was moved as rapidly as possible, and by December 15, fighters from the wing began orientation missions over Korea. The aircraft were equipped with six .50-caliber machine guns and flown by experienced pilots, many of whom had become combat aces during World War II. On December 17, 1950, the first clash of swept-wing fighters occurred between MiG-15s and F-86A Sabres. During the fight, Lt. Col. Bruce H. Hinton shot down a MiG-15 on patrol near the Yalu River, becoming the first F-86 pilot to shoot down a MiG in air-to-air combat.

The F-86's arrival drastically changed the air war over Korea. MiG-15 pilots became more cautious, and both sides developed new tactics for the new jet-versus-jet engagements. The Sabres continued to patrol the skies near the western Yalu River, which became known as "MiG Alley," and engage any aircraft that crossed over. The superior training of American pilots, combined with the abilities of the F-86, allowed the US Air Force to aggressively defend the skies over Korea during the length of the conflict. By the end of the Korean War, F-86 jets had destroyed almost eight hundred MiG-15s while suffering the loss of fewer than eighty of their own number. This incredible victory ratio cemented the F-86's legacy.

The F-86 would receive numerous upgrades to armament, engines, and control systems over the course of numerous models, including the F-86D, which featured a radar nose and rockets instead of machine guns. Various American-built F-86 models were used by the US Air Force and by air forces around the world; many more were produced under license in Canada, Japan, and Italy.

Although the F-86 flew combat only in the Korean War and was replaced in the US Air Force by the F-100 Super Sabre, it remained in active service with air national guards and foreign militaries for decades. The National Air and Space Museum's F-86 was part of the 4th Fighter-Interceptor Wing first sent to Korea in December 1950 and operated from numerous bases throughout the country before returning to the United States in June 1952. The aircraft remained in service with US Air National Guard units before it was transferred to the museum on January 25, 1962. Visitors can view the aircraft that helped pave the way for swept-wing fighters in the United States and throughout the world, and helped to forever change aerial combat.

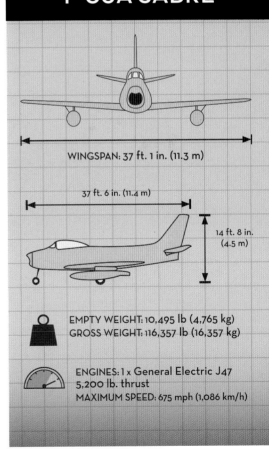

SPECIFICATIONS
NORTH AMERICAN F-86A SABRE

WINGSPAN: 37 ft. 1 in. (11.3 m)

37 ft. 6 in. (11.4 m)

14 ft. 8 in. (4.5 m)

EMPTY WEIGHT: 10,495 lb (4,765 kg)
GROSS WEIGHT: 116,357 lb (16,357 kg)

ENGINES: 1 x General Electric J47 5,200 lb. thrust
MAXIMUM SPEED: 675 mph (1,086 km/h)

BOEING 367-80

CHAPTER 19

BY F. ROBERT VAN DER LINDEN

On July 15, 1954, a graceful, swept-winged aircraft bedecked in brown and yellow paint and powered by four revolutionary new engines first took to the sky above Seattle. Built by the Boeing Aircraft Company, the 367-80, better known as the "Dash 80," would revolutionize commercial air transportation when its developed version entered service as the famous Boeing 707, America's first jet airliner.

Boeing became interested in jet transport designs following the initial success of the British de Havilland DH 106 Comet 1, the world's first jet airliner. With the Comet, Great Britain took the lead in civil jet propulsion. Hoping to leapfrog the United States' overwhelming domination of the piston-engine airliner market at the end of World War II, Lord John Moore-Brabazon's committee to develop the postwar aviation industry in Great Britain recommended several different aircraft configurations. One of these was for a sleek, jet-powered airliner seating forty-four passengers. With four de Havilland Ghost 50 turbojets buried in the roots of the slightly swept wings, the stylish all-metal DH 106 Comet first flew on July 27, 1949, with famed test pilot John Cunningham at the controls. In a series of demonstration flights, the Comet displayed a comfortable cruising speed of almost 450 miles per hour—100 miles per hour faster than contemporary conventional airliners.

Boeing restored the 367-80 and flew it from Seattle to Washington Dulles International Airport, where it was installed in the Boeing Aviation Hangar of the Smithsonian's Steven F. Udvar-Hazy Center in 2003.

TIMELINE
BOEING 367-80

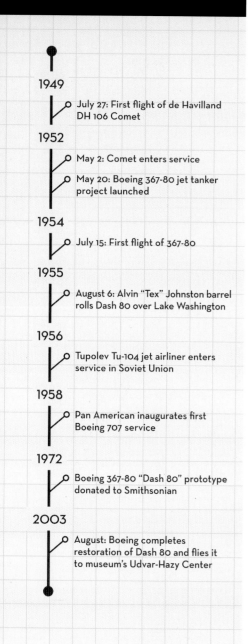

1949

○ July 27: First flight of de Havilland DH 106 Comet

1952

○ May 2: Comet enters service

○ May 20: Boeing 367-80 jet tanker project launched

1954

○ July 15: First flight of 367-80

1955

○ August 6: Alvin "Tex" Johnston barrel rolls Dash 80 over Lake Washington

1956

○ Tupolev Tu-104 jet airliner enters service in Soviet Union

1958

○ Pan American inaugurates first Boeing 707 service

1972

○ Boeing 367-80 "Dash 80" prototype donated to Smithsonian

2003

○ August: Boeing completes restoration of Dash 80 and flies it to museum's Udvar-Hazy Center

Exhaustive testing was the hallmark of de Havilland, and after minor modifications, the first Comet entered regular service with the British Overseas Airways Corporation (BOAC) on May 2, 1952, from London to Johannesburg. The Comet cut travel time by half on most of its routes while airlines rushed to place orders. Beginning in early 1953, however, a series of accidents showed that the new airliner required a wing modification, which was quickly made. Unfortunately, during the lengthy investigation, two more Comets were lost to unknown causes, forcing the grounding of all the aircraft. More thorough investigations revealed that the constant cycle of pressurization, particularly at higher altitudes, hastened the dangerous phenomenon of metal fatigue in the aircraft's structure, which led to a total failure of the fuselage due to cracks propagating from the square corners of the aircraft's top automatic direction finder (ADF) window. The investigation completed by de Havilland and the British government resulted in solutions that were quickly adopted throughout the international aviation community.

During this time, the Soviet Union jumped on the jetliner bandwagon with its own advanced airliner. Based on the excellent Tupolev Tu-16 medium bomber, Andrei Tupolev's Tu-104 featured a low-wing design powered by two large Mikulin AM-3-200 turbojet engines, installed in the wing root in a style similar to the Comet, and a fully pressurized cabin that could seat fifty passengers. The Tu-104 also featured a more pronounced wing sweep of 35 degrees, the result of German research during World War II.

By 1958 the new Comet 4, which was capable of crossing the Atlantic Ocean between London and New York, entered service, but it was too late. Boeing had designed its own airliner, the Model 367-80, which was considerably larger and faster than the Comet and incorporated all the design fixes learned from the de Havilland's trials.

Looking to develop a turbine-powered successor to the aging piston-engine C-97 military transport in service with the US Air Force, Boeing undertook a number of design studies to examine the possibility of creating a jet-powered military transport and tanker to complement the new generation of Boeing jet bombers entering service with the USAF. When the air force showed no interest, Boeing invested $16 million of its own capital to build a prototype jet transport in a gamble that airlines and the air force would buy it once the aircraft had proven itself. Initially, Boeing engineers based the jet transport on improved designs of the C-97, known at Boeing as the Model 367. By the time Boeing reached the eightieth iteration, the design bore no resemblance to the C-97, but for security reasons Boeing decided to let the jet project be known as the 367-80.

Work proceeded quickly after the formal start of the project on May 20, 1952. The 367-80 mated a large cabin based on the dimensions of the C-97 with the 35-degree swept-wing design based on the B-47 and B-52 but considerably stiffer and incorporating a pronounced dihedral. The wings were mounted low on the fuselage and incorporated high-speed and low-speed ailerons, as well as a sophisticated flap and spoiler system. Four Pratt & Whitney JT3 turbojet engines, each producing 10,000 pounds of thrust, were mounted on struts beneath the wings.

Upon the Dash 80's first flight on July 15, 1954 (Boeing's thirty-eighth anniversary), it was clear that the company had a winner. Significantly larger than the de Havilland Comet and flying 100 miles per hour faster, the new Boeing had a maximum range of more than 3,500 miles. As hoped, the air force bought 29 examples as a tanker/transport after convincing Boeing to widen the design by 12 inches. Satisfied, the air force designated it the KC-135A. A total of 732 would be built.

With the military order guaranteeing production, Boeing pitched a commercial version of the Dash 80, the Model 707, to airlines. The industry was impressed with the capabilities of the prototype 707, but never more so than at the Gold Cup hydroplane races held on Seattle's Lake Washington in August 1955. During the festivities, Boeing gathered several airline representatives to enjoy the competition and witness a fly-past of the new Dash 80. To the audience's intense delight—and Boeing's profound shock—test pilot Alvin "Tex" Johnston barrel-rolled the Dash 80 over the lake in full view of thousands of astonished spectators. Such a vivid display of the superior strength and performance of this new jet readily convinced the industry to buy the new airliner. It took years for Boeing president William Allen to forgive Johnston. After all, $16 million and Boeing's future were carried by that aircraft's wings—one mistake could have been fatal to Johnston and the company.

Above: The prototype British de Havilland Comet first flew in 1949 and was noted for its speed and smooth ride.

Opposite: The de Havilland DH 106 Comet, the world's first jet-powered airliner, revolutionized air travel when it entered service in 1952.

Below: In 1956, the Soviet Tupolev Tu-104 became the second jet airliner in service, serving short to medium routes continuously for Aeroflot.

In searching for a market, Boeing found a ready customer in Pan American World Airways' president. Juan Trippe had been spending much of his time searching for a suitable jet airliner to enable his pioneering company to maintain its leadership in international air travel. Pan Am was the first US airline to appreciate the speed and cost advantages of jet travel. Trippe overcame Boeing's resistance to widening the 707 design to seat six passengers in each seat row rather than five by placing an order for twenty 707s but also ordering twenty-five of competitor Douglas's DC-8, which had yet to fly but could accommodate six-abreast seating. Boeing relented, and the 707 was made 4 inches wider than the USAF's KC-135 to accommodate 160 passengers six-abreast. (The new 707 was also 17 feet longer than the Dash 80.) The wider fuselage became the standard design for all subsequent Boeing narrow-body airliners.

Although the de Havilland DH 106 Comet and the Tupolev Tu-104 entered service earlier, the Boeing 707 and Douglas DC-8 were bigger and faster, had greater range, and were more profitable to fly. On October 26, 1958, Pan American ushered in the jet age in the United States when it opened international service with the Boeing 707. National Airlines inaugurated domestic jet service two months later using a 707-120 borrowed from Pan Am. American Airlines flew the first domestic 707 jet service with its own aircraft in January 1959. Subsequent nonstop flights between New York and San Francisco took only five hours—three hours less than by the piston-engine DC-7. The flight was also almost 40 percent faster and 25 percent cheaper than a piston-engine flight. The consequent surge of traffic was substantial.

To improve stability and prevent over-rotation on takeoff, 707s were fitted with a ventral fin under the tail and a heightened rudder. Many other modifications and improvements were made over the course of the 707's distinguished career.

Due to high fuel consumption, early jet transports were known for their speed but not their range. The first generation of pure turbojet engines was quite thirsty and inefficient, leaving behind a thick trail of smoke—unburned hydrocarbons. Working closely with the engine manufacturers, particularly Pratt & Whitney, Boeing developed a larger, longer-ranged version of the 707, known as the 707-300 Intercontinental. With room for 189 passengers and fitted with a larger wing and tail, as well as more powerful Pratt & Whitney JT4 turbojets, the Intercontinental had a range of over 4,000 miles and could cross the Atlantic with ease. The 707-300B was the most successful version, incorporating more wing area and additional flaps, extra fuel tanks, and most significantly, more efficient Pratt & Whitney JT-3D turbofan engines.

Opposite left: The Dash 80 cockpit was a utilitarian design typical of the era and featured a flight engineer's station from which the aircraft's systems were monitored.

Opposite right: For many years, Boeing used the Dash 80 to test new equipment and features, including the Pratt & Whitney JT3B turbofan engines shown here.

Below left: The graceful Dash 80 featured highly swept wings and podded turbojets engines, two technologies derived from World War II German research.

Below right: Boeing test pilot "Tex" Johnson made headlines when he rolled the 367-80 "Dash 80" over Lake Washington, Seattle, during the 1955 Gold Cup hydroplane races.

The US Air Force watched 707 developments closely. While it maintained a huge fleet of smaller KC-135 jet tankers, it needed the larger 707 for numerous missions, such as airborne early-warning and electronic countermeasures and, most importantly, transporting the president of the United States and senior leaders of government. Technically designated VC-137s, the first of several USAF 707s were delivered in 1959. Soon, the "Air Force One" designation was applied to any aircraft carrying the president. The first three of these were later re-engined with JT3Ds, while two special 707-353Bs were ordered in 1961 specifically for the use of the president and fitted with the latest communications equipment and interior features. These latter two aircraft, in particular, came to be known as "Air Force One" in the public's mind.

Most remarkably, because of its excellent design and the smoothness of its jet engines, the airframe life of the 707 exceeded those of all of its predecessors. As jetliner design matured in the 1960s, the 707 remained technologically current and was not replaced on its primary routes until a new generation of long-range jetliners—including the Boeing 747 and 767, the McDonnell Douglas DC-10, and the Lockheed L-1011—introduced more efficient and cleaner engines in the 1970s and subsequent decades. All told, Boeing built 855 707s, of which 725 were bought by airlines around the world.

Having launched the Boeing Company into the commercial jet age, the Dash 80 soldiered on as a highly successful experimental aircraft. Until its retirement in 1972, it was used to test numerous advanced systems, many of which were incorporated into later generations of jet transports. At one point, it carried three different engine types in its four nacelles. While serving as a test bed for the new 727, the Dash 80 was briefly equipped with a fifth engine mounted on the rear fuselage. Engineers also modified the wing in planform and contour to study the effects of different airfoil shapes, and numerous flap configurations were fitted, including a highly sophisticated system of "blown" flaps over which engine exhaust was redirected to increase lift at low speeds. Fin height and horizontal stabilizer width were later increased, and at one point a special multiple-wheel low-pressure landing gear system was tested to study the feasibility of operating heavy military transports from unprepared landing fields.

After a long and distinguished career, the Boeing 367-80 was finally retired and an example donated to the Smithsonian in 1972. It was painstakingly restored by the Boeing Company during the 1990s and flown in August 2003 to the National Air and Space Museum's new Steven F. Udvar-Hazy Center, where it sits proudly on display, the vanguard of the Jet Age.

SPECIFICATIONS
BOEING 367-80

WINGSPAN: 129 ft. 8 in. (39.5 m)

127 ft. 10 in. (39 m)

38 ft. (11.6 m)

EMPTY WEIGHT: 92,120 lb. (41,784 kg)

ENGINES: 4 x Pratt & Whitney JT3 turbojets, 10,000 lbf (44,482 kN) thrust each
MAXIMUM SPEED: 550 mph (885 km/h)

McDONNELL F-4S-44 PHANTOM II

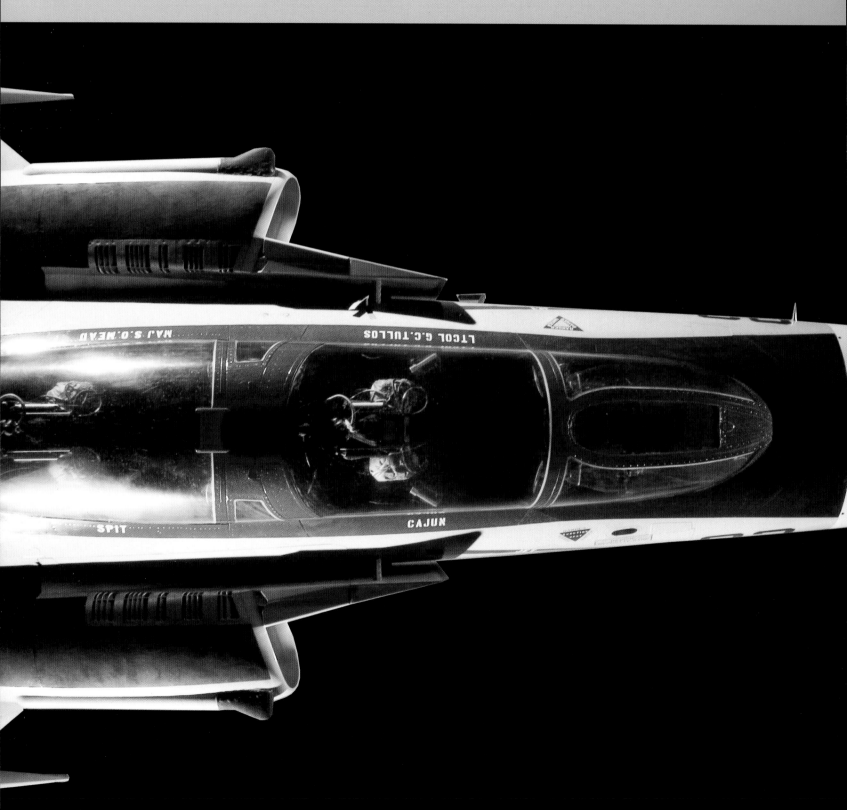

CHAPTER 20

BY ALEX M SPENCER

In August 1953, engineers at the McDonnell Aircraft Company began design of a new long-range interceptor aircraft equipped with the new and powerful General Electric J79 engine. At that time there was no demand or customer for the aircraft. From the perspective of the management at McDonnell, however, the new aircraft would fill the US Navy's need for a long-range standoff fighter. The final configuration of this aircraft, the F-4 Phantom II, became one of the most produced and longest serving jet fighters in history.

The F-4 design began as a single-place, long-range attack aircraft. Working with the navy, McDonnell engineers reconfigured the design of the company's F3H-G Demon into a new delta-wing configuration. In November 1955, a full-scale model of the new aircraft was presented to the navy. Though there was no military requirement for such an airplane, the navy was interested in its potential as a standoff fleet air-defense aircraft using the long-range Sparrow air-to-air missile. This required one significant modification to the McDonnell preliminary design: the addition of a second cockpit to accommodate a weapons systems operator.

During the aircraft's development, wind-tunnel testing uncovered a number of significant stability problems in the slower flight regimes. To correct these teething troubles,

An overhead view of the F-4 Phantom on display at the Steven F. Udvar-Hazy Center

1953
○ McDonnell begins new design of J79 powered interceptor

1955
○ November: Design presented to US Navy

1958
○ May 27: First flight of McDonnell F4H-1 Phantom II (F-4A)

1963
○ February 8: F-4C US Air Force version authorized

1964
○ August 5: US Navy F-4Bs strike North Vietnamese targets after Gulf of Tonkin incident

1965
○ July 10: First Phantom victory when Air Force F-4C downs MiG-17

1968
○ July: Delivery of first Phantoms for Royal Air Force and Fleet Air Arm

1969
○ March: F-4s delivered to Israeli Air Force

1972
○ June 21: F-4 #157307 destroys MiG-21

1973
○ October 6: F-4 plays key role in Yom Kippur War

1979
○ Last Phantom II produced

1986
○ October: Last carrier landing of F-4 for US Navy

1988
○ November 28: F-4 #157307 transferred to NASM

a 12-degree dihedral was added to the wingtips and a 23-degree anhedral angle was given to the stabilator. These modifications gave the Phantom its unique appearance.

In addition, modifications to the J79 engines, air intakes, electronics, and canopy resulted in the final aircraft configuration designated the F4H-1, the navy's first Mach 2 carrier-based aircraft and the first American all-missile fighter whose primary mission was as an all-weather fleet air-defense aircraft. The intent of its mission was to have the aircraft loiter 250 miles from a task force and shoot down at long range with the Sparrow missile any aircraft that posed a threat. Although it was seen primarily as a fighter, the aircraft's design retained ground attack and reconnaissance capabilities.

The F4H-1 made its first flight on May 27, 1958, and in December, McDonnell was awarded a limited production contract. One year later, the redesignated F-4B Phantom II joined the fleet and was assigned to Fighting Squadron 121. The Phantom was quickly qualified for both land and sea operations.

With the election of President John F. Kennedy in 1960 and the subsequent appointment of Robert McNamara as secretary of defense, the administration changed the emphasis of the country's strategic priorities. In March 1961, Kennedy presented his vision to the US Congress. He wanted to deemphasize the nuclear deterrent and massive retaliation and emphasize a policy of "flexible response" and increase the nation's "limited warfare" capabilities. This new strategy required the US Air Force to expand its conventional weapons capabilities. In addition, McNamara wanted to reduce defense spending by encouraging the armed forces to adopt commonality with their equipment. The performance and versatility of the F-4's airframe could provide the air force with a conventionally armed aircraft capable of serving as a fighter, tactical fighter, and reconnaissance aircraft. In January 1962, the air force began testing the aircraft's capabilities, and within two months the Department of Defense announced that the Phantom would become Tactical Air Command's standard fighter and reconnaissance aircraft. Production of the Phantom for the USAF as the F-4C was authorized on February 8, 1963.

Between 1959 and 1969, the F-4 and its derivatives established many altitude and speed records. However, the aircraft came to the fore with its extensive use by the US Navy, Marines, and Air Force in the Vietnam War. On August 5, 1964, navy F-4Bs from squadrons VF-142 and VF-143 of the USS *Constellation* flew air cover for the first aerial retaliatory strike following the Gulf of Tonkin Incident. The aircrews flying the Phantom distinguished themselves and the aircraft by flying tens of thousands of air strikes and in air-to-air combat. US Air Force Capts. T. S. Roberts and R. C. Anderson were credited with the first aerial victory of the war, shooting down a MiG-17 on July 10, 1965. All of the Americans credited with becoming aces during the war were pilots or weapons system operators flying the Phantom. They made the first attacks utilizing laser-guided munitions. However, the Phantom crews also paid a heavy price for their involvement—the air force lost 525 aircraft, the navy 71, and the marines 72. The majority of these losses were from antiaircraft missiles and guns.

The McDonnell Douglas F-4 found its way into the international market with an overseas order made by the British in September 1964, though the order did not reach Britain until July 1968. The aircraft served both the Royal Air Force and Royal Navy's Fleet Air Arm (FAA). The principal point on which the British version differed from the American was the Rolls-Royce Spey engines that replaced the original J79s. The Phantom was the FAA's first truly supersonic aircraft and served with 892 Squadron on HMS *Ark Royal* from June 1970 to November 1978. The aircraft was withdrawn from the FAA when the Royal Navy abandoned the use of fleet-sized aircraft carriers at the end of 1978.

The Phantom also saw extensive combat service in the Middle East with the Israeli Air Force (IAF). The Israelis expressed interest in obtaining the F-4 as early as 1965, but the United States was unwilling to sell its most advanced fighter overseas. At that time, the IAF was equipped primarily with French Dassault Mirage jet fighters, but following the Six Day War in 1967 and an assault on the Beirut airport by commandos in 1968, France embargoed all sales of military equipment to Israel. Once again Israel requested the F-4, and in January 1968 the Johnson administration agreed to the sale of forty-four aircraft. Deliveries and the training of Israeli pilots began in March 1969. Until the IAF F-4s became operational in May 1971, Israel was locked in the War of Attrition with neighboring countries, marked by continual strike and counterstrike duels.

With the IAF F-4s the balance soon tipped in favor of Israel. With their long-range missile capability, the aircraft could stand off at long distances and destroy Egyptian and Syrian surface-to-air antiaircraft batteries. The IAF quickly demonstrated its improved situation by making unimpeded supersonic passes over the Egyptian capital of Cairo.

On October 6, 1973, the combined armies of Egypt and Syria launched an all-out assault on Israel. At the start of the Yom Kippur War, the 127 IAF F-4 Phantoms were the centerpieces of the IAF's aerial operations and conducted the majority of IAF air combat, ground support, and reconnaissance missions. Armed with the F-4, the IAF soon

Above: This view shows the front cockpit of the National Air and Space Museum's F-4. The addition of long-range air-to-air missiles to the armament of the preliminary design required one significant modification: the addition of a second cockpit for a weapons systems operator.

Opposite top left: The Smithsonian's National Air and Space Museum F-4 Phantom is on display at the Udvar-Hazy Center. The aircraft ended its career as a modified S variant with Marine F232.

Opposite top right: The National Air and Space Museum's F-4 flew combat operations over Vietnam from May to June 1972 off of the deck of the USS *Saratoga*.

Opposite bottom: Commander S. C. Flynn scored a MiG kill on June 21, 1972, while flying F-4 No. 106, which would become the National Air and Space Museum's F-4 Phantom. Flynn, however, painted the victory marking on F-4 No. 102, his regularly assigned aircraft.

The MiG-21 was the F-4's primary adversary during the Vietnam War was the MiG-21. This former Soviet aircraft is part of the National Air and Space Museum's collection at the Udvar-Hazy Center. Its prior history is unknown.

established air superiority by shooting down over 277 Arab Soviet-supplied MiG-17, MiG-19, and MiG-21 fighters.

The F-4 Phantom would serve with nine additional air forces, namely Egypt, Iran, South Korea, Spain, Australia, Japan, Greece, Turkey, and West Germany. Specially designed for Japan, the F-4EJ variant dispensed with most of the offensive systems and was fitted with advanced tail-warning radar and air-to-air guided missiles. (Japan also ordered the RF-4EJ, a fully unarmed reconnaissance version.) Japan assembled 11 of its F-4s in Japan and built 126 under license.

By the time Phantom production ceased in 1979, 5,195 had been built. The last navy F-4 made its final "trap" (carrier landing) aboard the USS *America* in October 1986. The German Air Force was the last air force to use the aircraft operationally, with several Phantoms at Holloman Air Force Base serving as training platforms for their fighter pilots until 2013. The Phantom proved to be a versatile aircraft design, and over the course of its production thirty-seven distinct types were manufactured.

The National Air and Space Museum's F-4S-44 McDonnell Douglas Phantom II, Bureau Number (BuNo) 157307, was accepted by the navy on December 18, 1970. By June 22, 1971, it was assigned to Fighting Squadron 31 (VF-31) stationed at the Naval Air Station at Oceana, Virginia. Early in 1972, VF-31 (along with F-4 BuNo 157307) went aboard USS *Saratoga* and by April were en route to the western Pacific for duty in the Vietnam War. On May 18, 1972, the squadron started combat operations on Yankee Station, off the coast of Vietnam. While on a flight on June 21, 1972, its last day on station, F-4 BuNo 157307 (Squadron 106) made its mark. It was launched that day on a MiGCAP (MiG Combat Air Patrol) with VF-31's executive officer, Cdr. S. C. Flynn, as pilot and Lt. W. H. John as the radar intercept officer (RIO). This was not their regularly assigned airplane. They spotted three MiGs and in the ensuing engagement shot down one MiG-21 with a Sidewinder missile (AIM-9). Their action marked a first for the *Saratoga* air wing and for an East Coast fighter squadron.

After the kill, the Phantom's tour in Vietnam was expanded to include support missions for B-52 raids on Hanoi and Haiphong. VF-31 completed its deployment to Southeast Asia early in 1973 and returned to its home port at Oceana.

The Phantom remained assigned to VF-31 until September 12, 1975, when it was transferred to VF-33. After a series of deployments aboard USS *Independence*, it was assigned to VF-74 on May 6, 1977, also based at Oceana. It left VF-74 on September 17, 1979, for VF-103, then went to VF-171 on October 21, 1981. On April 8, 1983, F-4J BuNo 157307 was inducted into the Naval Air Rework Facility at North Island, California, for conversion from a J model to an S model.

The S conversion was an extensive modernization and service-life extension overhaul for 250 F-4Js. It consisted mainly of modernizing the hydraulics, electronics, and wiring, and later included installation of leading-edge maneuvering slats (like those on the F-4Es and F-4Fs), radar homing and warning (RHAW) antenna, and formation tape lights on the fuselage and vertical tail.

When conversion was completed on December 27, 1983, F-4S 157307 joined Marine Fighter Attack Training Squadron (VMFAT) 101, stationed at Marine Corps Air Station (MCAS) Yuma, Arizona. It remained there until May 11, 1987, when it was transferred to VFMA-232, Honolulu on its last squadron duty. On November 28, 1988, it left the marines for the National Air and Space Museum, at which time it had amassed a total of 5,075 hours flight time with 6,804 landings (1,337 were arrested), and 1,163 catapult shots off carrier decks.

Below: McDonnell Aircraft celebrated the production of the 5,000th F-4 with a special paint scheme that included the flags of the nations that operated the aircraft.

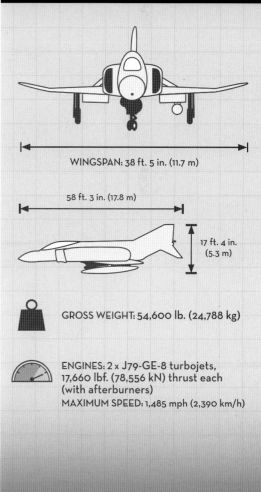

SPECIFICATIONS
MCDONNELL F-4S-44 PHANTOM II

WINGSPAN: 38 ft. 5 in. (11.7 m)

58 ft. 3 in. (17.8 m)

17 ft. 4 in. (5.3 m)

GROSS WEIGHT: 54,600 lb. (24,788 kg)

ENGINES: 2 x J79-GE-8 turbojets, 17,660 lbf. (78,556 kN) thrust each (with afterburners)
MAXIMUM SPEED: 1,485 mph (2,390 km/h)

NORTH AMERICAN X-15

CHAPTER 21

BY JOHN D. ANDERSON JR.

The time: early morning on October 3, 1967. The location: about 40,000 feet above Edwards Air Force Base in the Mojave Desert in Southern California. The action: inside the cockpit of a pioneering hypersonic test airplane mounted under the right wing of a B-52 bomber, test pilot William "Pete" Knight is carefully checking out the airplane's systems; at his command, the pilot and airplane are released from the B-52 and Knight starts the rocket engine.

The airplane is the X-15 research vehicle. Responding to 57,000 pounds of thrust from the rocket engine, it accelerates rapidly. On this flight, the airplane ultimately accelerates to a Mach number of 6.7, the highest achieved by any of the 199 flights of the X-15.

Four years earlier, on August 22, 1963, test pilot Joe Walker had flown the X-15 to an altitude of 354,200 feet, the highest achieved by the airplane. These two records still hold—the X-15 remains the fastest and highest-flying piloted, powered, winged airplane in the history of aeronautics. The first airplane to fly at hypersonic speeds (faster than Mach 5), it embodied the mantra that has driven the progress of airplane design since the first flight of the Wright Flyer: faster and higher.

Three X-15s were built by its manufacturer, North American Aviation (NAA). The first now hangs with distinction in the Boeing Milestones of Flight Hall at the National Air and Space Museum.

This head-on view provides a good close-up of the X-15 hanging in the Smithsonian National Air and Space Museum.

TIMELINE
NORTH AMERICAN X-15

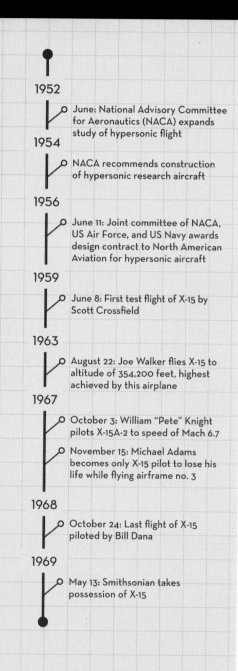

1952

June: National Advisory Committee for Aeronautics (NACA) expands study of hypersonic flight

1954

NACA recommends construction of hypersonic research aircraft

1956

June 11: Joint committee of NACA, US Air Force, and US Navy awards design contract to North American Aviation for hypersonic aircraft

1959

June 8: First test flight of X-15 by Scott Crossfield

1963

August 22: Joe Walker flies X-15 to altitude of 354,200 feet, highest achieved by this airplane

1967

October 3: William "Pete" Knight pilots X-15A-2 to speed of Mach 6.7

November 15: Michael Adams becomes only X-15 pilot to lose his life while flying airframe no. 3

1968

October 24: Last flight of X-15 piloted by Bill Dana

1969

May 13: Smithsonian takes possession of X-15

The X-15 was purely and simply a research airplane, stemming from a philosophy initiated by the pioneering Bell X-1, the first airplane to break the speed of sound when, piloted by Capt. Chuck Yeager, it achieved Mach 1.07 on October 14, 1947. The X-1 was designed in the mid-1940s to fill a vital aeronautical knowledge gap that existed at the time, namely what happens in the speed range just below and just above Mach 1 (transonic flight). In the 1940s, no wind-tunnel data or aerodynamic theory existed to allow the accurate study and prediction of aerodynamic behavior in the transonic range. Ultimately, the aeronautical community accepted that the only way to obtain such data was to build an actual airplane and fly it at transonic speeds. It was for this purpose only that Bell Aircraft designed the X-1, thus setting the precedent for purely research airplanes, more popularly known as X-planes. (The series is up to the X-51, an unmanned hypersonic vehicle powered by an air-breathing supersonic combustion ramjet engine.)

The first hypersonic vehicles in flight were missiles, not airplanes. On February 24, 1949, a WAC Corporal rocket mounted on top of a captured German V-2 boost vehicle was fired from the White Sands Proving Ground in New Mexico, reaching an altitude of 244 miles and a velocity of 5,150 miles per hour. After nosing over, the WAC Corporal careened back into the atmosphere at more than 5,000 miles per hour, becoming the first object of human origin to achieve hypersonic flight. During this same period, a hypersonic wind tunnel capable of Mach 7 with an 11-by-11-inch cross-section went into operation on November 26, 1947, the brainchild of National Advisory Committee for Aeronautics (NACA) Langley researcher John Becker. For three years following its first run, this wind tunnel was the only hypersonic wind tunnel in the United States.

The state of knowledge of hypersonic aerodynamics and the problems of hypersonic flight were embryonic at this time, and the same type of knowledge gap that prompted the existence of the Bell X-1 in transonic and supersonic flight now existed in the hypersonic flight regime. In this flight regime, several aerodynamic phenomena become important that are not so important at supersonic speeds and thus were not faced by the Bell X-1. The most serious of these is severe aerodynamic heating caused by strong shock waves and intense surface skin friction characteristic of high-speed flow over a flight vehicle. The intense aerodynamic heating greatly affects the external and internal structure of the airplane, creating an aero-thermodynamic/structural interaction that dominates the design characteristics of all hypersonic vehicles.

In the 1950s, no accurate hypersonic aerodynamic theories or workhorse hypersonic wind tunnels existed to provide design data for hypersonic airplanes; once again, flying an actual experimental airplane was the only viable option to probe the mysteries of this knowledge gap. In June 1952, the NACA Committee on Aerodynamics recommended that the NACA expand its efforts to study the problems of hypersonic manned and unmanned flight, covering the Mach number range from 4 to 10. Two years later, the same committee recommended the construction of a hypersonic research airplane. Thus, the X-15 was born.

The NACA played a strong role in the conceptual design of the X-15, and a team at the NACA Langley Aeronautical Laboratory in Hampton, Virginia, under the direction of Becker, defined the early requirements for the airplane that optimized research into the related fields of high-temperature aerodynamics and high-temperature structures. In November 1954, the NACA, air force, and navy signed a memorandum of understanding that gave NACA responsibility for the technical direction of the X-15 research project, with the air force and navy providing money and flight research facilities and pilots. The air force was

Top left: The ball nose reduced aerodynamic heating and indicated the attack and yaw angles. The forward control jets are also visible. Top right: The X-15 canopy has what's referred to as a hypersonic aerodynamic configuration. Bottom: This detail view shows the control jets on the X-15 fuselage.

Above: X-15-2 was photographed just after a launch from its B-52 "mothership" sometime in the early 1960s.

Opposite top: The X-15s were carried aloft in a B-52 "mothership" to save fuel by not having to take off from the ground.

Opposite bottom: Hollywood thought the X-15 fascinating enough to feature it in a 1961 motion picture costarring Charles Bronson and Mary Tyler Moore and narrated by James Stewart. The film was intercut with stock NASA footage.

Below: Neil Armstrong, more famous for being the first man to set foot on the moon, was an X-15 test pilot for seven flights from 1960 to 1962. He's seen here with X-15-1, now in the National Air and Space Museum.

authorized to issue invitations to twelve prospective aircraft contractors with experience in the development of high-performance airplanes to participate in the design competition for a Mach 7 research aircraft keying on NACA's conceptual design. Only four companies responded: North American, Douglas, Bell, and Republic. A joint committee representing the NACA, air force, and navy ultimately chose the NAA design, and the company was awarded a contract on June 11, 1956. Funding eventually amounted to $43 million for three X-15 aircraft. In September of the same year, Reaction Motors was contracted to provide the rocket engines for the X-15s.

Previously, no airplane had even come close to Mach 5. The Bell X-1 had achieved Mach 1, the Douglas D-558-2 Skyrocket had flown at Mach 2, and the Bell X-2 had reached Mach 3.2. To approve the development of an airplane that could fly at Mach 7 was truly visionary. Normally, engineers study previous incarnations of the new airplane they want to build, innovating from successful earlier designs. But with the X-15, no "before" design existed. The team had to start from scratch.

In order to hold up under the extreme aerodynamic heating, the external structure of the X-15 was made from Inconel X, an alloy that maintains its strength above 2,000 degrees Fahrenheit. The internal structure was mainly fabricated from titanium. Although the airplane's configuration is generally sleek and streamlined, as might be expected for a very high-speed airplane, the wings' leading edges are rounded (rather than sharp), and the nose has a spherical tip (rather than pointed). These blunt shapes may seem counterintuitive, but it has been known since 1953 that such shapes actually reduce aerodynamic heating more effectively than sharp shapes—the blunt nose of the Apollo lunar return capsule is an extreme case in point.

The X-15's short, stubby wingspan is only 22 feet, and the airfoil section is an NACA 66005 symmetric laminar flow airfoil. The horizontal tail is tilted down from the fuselage, while the upper vertical tail looks like most others except that the airfoil is wedge shaped with a blunt trailing edge, a new design feature for enhancing the directional stability of the airplane at high Mach numbers. There are no ailerons on the wing; roll control is provided by deflecting differentially the right and left sections of the horizontal tail. Also, the horizontal tail has no elevators; instead the entire right and left sections deflect in the same direction together to provide pitch control. These aerodynamic controls are for use in the sensible parts of the atmosphere—at the outer edge of the atmosphere, where air density is too low for aerodynamic control, small jet thrusters control the airplane.

The X-15's Reaction Motors XLR-99 engine is capable of generating 57,000 pounds of thrust on fuel of anhydrous ammonia, with liquid oxygen as the oxidizer.

Like its predecessors, the X-15 was flown out of Edwards Air Force Base in California, the location of the US Air Force Test Pilot School and the only installation with the support equipment and personnel to handle the research test flights. Moreover, because of the X-15's high landing speed, Edwards had the only "runway" long enough for landing the airplane—essentially the whole expanse of Rogers Dry Lake. The new airplane, like the X-1 and X-2 before it, was carried aloft in a B-52 "mothership" to save fuel by not having to take off from the ground; the X-15 pilot applied the thrust at an altitude where the air density was low (and, hence, so was aerodynamic drag). However, the X-15 still had to carry sufficient fuel to allow the high thrust to operate long enough to accelerate to the speeds and altitudes needed to perform the mission.

What was it like for an X-15 research test pilot to fly into unknown areas of speed and altitude? He arrives early in the morning, a good time for flight, since the desert winds and temperature are lower at that time. He goes to the physiological van at Edwards, and there he puts on his pressure suit. He walks across the ramp to the airplanes, the B-52 and the X-15 that is mated to its right wing. He climbs a large ladder to a platform next to the X-15 and then enters the small X-15 cockpit. After the test pilot and the B-52 crew go through preflight checklists, the B-52 takes off with the X-15 and its pilot captive under the bomber's wing. At 40,000 feet, the X-15 is launched, rolls slightly to the right, and levels out, and the pilot ignites the rocket engine.

The X-15 rapidly accelerates away from the B-52 and is on its way for the day's prescribed test flight. This might involve a speed or high-altitude test, or it might simply be an investigation of stability and control performance at hypersonic speeds. After the fuel and oxidizer are used up, the thrust drops to zero, and the airplane starts to fall back to Earth. In a controlled glide, the X-15 decelerates as it enters the denser parts of the atmosphere, finally approaching the landing site at 200 miles per hour, where it touches down on a tail skid and then a nose wheel. The entire length of flight, from drop to touchdown on the desert floor, usually lasts ten minutes.

The first test flight took place on June 8, 1959, with Scott Crossfield at the controls. Dropped from the B-52 from an altitude of 37,550 feet, he completed an unpowered flight lasting only four minutes and fifty-seven seconds; its purpose was to simply test the X-15's stability, control, and basic subsonic flight characteristics as it glided to the ground. A total of 198 more X-15 flights would follow, the last taking place on October 24, 1968, piloted by Bill Dana. On this last flight, Dana reached a Mach number of 5.38 and an altitude of 255,000 feet. In an effort to round out the X-15 program, a 200th flight was planned later that year, but on the prescribed day, bad weather—including snow—set in, and the mission was scrubbed. Because the funds for the program had been exhausted, no other flight took place.

From the first to the last flight, the X-15 was flown by a total of twelve highly qualified test pilots. Crossfield had been hired by North American to carry out the early part of the program associated with the required contractor flights before formal acceptance of the airplane by the air force. He flew the X-15 fourteen times, reaching the relatively moderate

Mach number of 2.97 and an altitude of 86,116 feet. During the early part of the X-15 program, the designated XLR-99 rocket engine was not available, and in order to move the X-15 program ahead, the airplane was powered by two XLR-11 engines with a total thrust of 12,000 pounds. Only after the twenty-fifth flight did the XLR-99, with its thrust of 57,000 pounds, become available.

Robert Rushworth was the workhorse test pilot for the X-15, with thirty-four flights. He achieved a Mach number of 6.06 and an altitude of 285,000 feet. Neil Armstrong piloted seven flights, achieving Mach 5.74 and an altitude of 201,500 feet. (Armstrong left prematurely to join the space program.) In addition to Pete Knight and Joe Walker, who had record-breaking flights (Mach 6.7 and 354,200 feet, respectively), other X-15 pilots were Robert White, Forest Petersen, John McKay, Joe Engle, Milt Thompson, and Michael Adams.

All pilots in the X-15 program were the cream of the crop: highly experienced and dedicated professionals venturing into a totally unknown regime of flight who were as finely tuned and technologically sophisticated as the machine itself. They set speed and altitude records for a manned airplane that still stand today, and they pioneered new piloting techniques for hypersonic aircraft that were not only adapted for the Space Shuttle but will continue to be used for future manned hypersonic aircraft. On almost all X-15 flights, some type of problem occurred, and the effective ways in which the test pilots dealt with these problems are as much responsible for the success of the program as the design of the vehicle and overall effective operation of the program.

One fatal accident occurred during the X-15 flight-test program. On November 15, 1967, Michael Adams, veteran pilot with six previous X-15 flights, entered the aircraft for a flight to evaluate a guidance display and to conduct several experiments. He had spent more than twenty-one hours practicing the flight specifics on the simulator. The drop at about 10:00 a.m. and 45,000 feet was normal. After burnout, Adams soared upward, achieving a peak altitude of 266,000 feet. Then a problem set in: his yaw had drifted to 15 degrees, but he was unaware because his instrument was inadvertently set to show pitch, not yaw. About fifteen seconds later, the airplane yawed wildly and then entered a spin. When Adams finally recovered, the flight data showed that he was yawed 90 degrees, flying

An X-15-2 lands in the Mojave Desert with its nose wheel and rear skids extended.

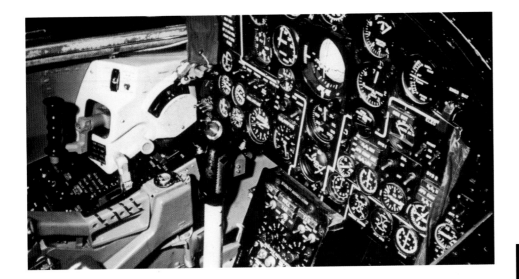

This view of the X-15 cockpit the Smithsonian National Air and Space Museum shows the unique control sticks used to control the whole flight regime.

upside down, and descending at supersonic speed. Due to a problem with the flight control system, Adams was unable to correct this attitude. The loads on the airplane built up beyond the structural limits, and the X-15 broke up at approximately 62,000 feet and about 3,800 feet per second, crashing to the desert floor near Johannesburg, California. Adams's death underscored the dangers of testing new aircraft in previously untested regions of flight and of flying experiments in which certain data measuring instruments may cause problems. Considering the pioneering nature of the X-15 design, the unknown flight characteristics of the hypersonic flight regime, and the very nature of flight testing at the extremes of speed and altitude, it is remarkable that all other 198 flights got home safely

All X-15 design goals were met during the flight-test program, and some were surpassed. The design maximum altitude and Mach number were both reached. The hypersonic research data obtained provided a rich database that confirmed the viability of hypersonic wind-tunnel data as well as the usefulness of the limited theoretical analyses of hypersonic aerodynamics available at the time. The airplane proved a successful hypersonic vehicle, and the pilots performed admirably over an almost ten-year period. The X-15 program ended only when funding was exhausted and research experiments no longer justified the associated costs of the flights.

As early as 1962, the Smithsonian Institution requested an X-15 airplane for eventual display in Washington, D.C. The first X-15 was received on May 13, 1969, in what was then known as Silver-Hill (now the Garber Facility). It was moved to the Smithsonian's Arts and Industries Building in June 1969 and placed near the 1903 Wright Flyer. The Arts and Industries building served as the National Air and Space Museum at that time. After being loaned to the Federal Aviation Administration and then to the NASA Flight Research Center for display, it returned to the Smithsonian to be installed for the opening of the new National Air and Space Museum on the Mall in Washington, on July 1, 1976. It hangs there today in the Milestones of Flight Hall.

The future of practical, environmentally safe, and economically feasible hypersonic manned flight still lies before us, and when that happens, the X-15 will indeed be the "Wright Flyer" of its kind.

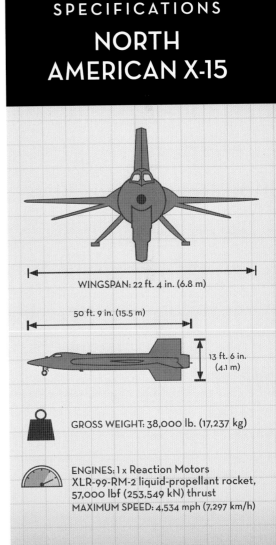

SPECIFICATIONS
NORTH AMERICAN X-15

WINGSPAN: 22 ft. 4 in. (6.8 m)

50 ft. 9 in. (15.5 m)

13 ft. 6 in. (4.1 m)

GROSS WEIGHT: 38,000 lb. (17,237 kg)

ENGINES: 1 x Reaction Motors XLR-99-RM-2 liquid-propellant rocket, 57,000 lbf (253,549 kN) thrust
MAXIMUM SPEED: 4,534 mph (7,297 km/h)

CHAPTER 22

BY RUSSELL E. LEE

Danish pilot Harald Jensen said in 1959, "We already have a 1,000-kilometer club. Now all we need is some members."

On July 31, 1964, Alvin H. Parker flew from his hometown of Odessa, Texas, at the controls of the Sisu 1A sailplane and, in so doing, shattered a symbolic and psychological barrier that had defeated sailplane pilots for years: the first flight to exceed 1,000 kilometers. In his soaring anthology *On Quiet Wings* (1972), Joseph Lincoln described the 1,000-kilometer flight as "the soaring pilot's four-minute mile."

In addition to the record 1,042 kilometers (647 miles) flown in 1964, Parker also set one earlier and one later world record flying the Sisu (pronounced see-sue): the first when he achieved a declared-goal distance flight of 487 miles in 1963 and the second when he smashed that record flying another declared goal of 578 miles from Odessa in 1969.

Parker was not the only pilot to achieve success in a Sisu glider. John Ryan (1962), Dean Svec (1965), and A. J. Smith (1967) all won the United States National Soaring Championships flying Sisus. In 1967, Bill Ivans set a national speed record at El Mirage, California, by skimming across the desert in a Sisu at 135 kilometers per hour (84 miles per hour) over a 100-kilometer (62-mile) triangular course.

Al Parker's Sisu 1A, on display at the Smithsonian National Air and Space Museum's Udvar-Hazy Center.

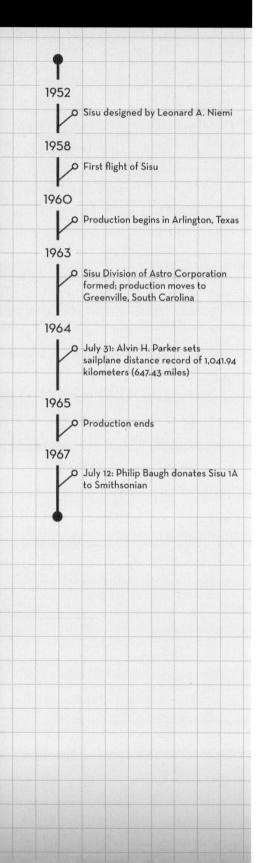

TIMELINE
ARLINGTON SISU 1A

1952
Sisu designed by Leonard A. Niemi

1958
First flight of Sisu

1960
Production begins in Arlington, Texas

1963
Sisu Division of Astro Corporation formed; production moves to Greenville, South Carolina

1964
July 31: Alvin H. Parker sets sailplane distance record of 1,041.94 kilometers (647.43 miles)

1965
Production ends

1967
July 12: Philip Baugh donates Sisu 1A to Smithsonian

Before any of these flights, Richard Johnson's 1951 world distance record flight of 547 miles in the Ross-Johnson RJ-5 sailplane had demonstrated the superiority of the low-speed laminar flow airfoils fitted to the aircraft. This advanced wing design greatly smoothed the airflow around it, thus reducing drag significantly. Harlan Ross had built the RJ-5, and Richard Johnson had refined it with guidance from Dr. August Raspet at the Mississippi State University.

The success of the RJ-5 marked the beginning of a profound change in the philosophy of high-performance sailplane design. The innovations incorporated in the RJ-5 inspired Leonard A. Niemi to design the Sisu in 1952. Niemi had spent four years learning the aircraft mechanics trade while attending a technical high school in Buffalo, New York, followed by another six years working for the Bell Aircraft Corporation and Curtiss-Wright. After a stint in the US Army, he attended the University of Michigan and earned a BS in aeronautical engineering. He went to Bell Helicopter and spent four years performing stress analysis.

Niemi's ancestors were Finns, which probably led him to name his design the Sisu. It is a popular Finnish word with no precise English equivalent that describes fundamental characteristics of native Finns—their strength, perseverance, and strong will. Aini Rajanen wrote in *Of Finnish Ways* (1984) that *sisu* is "a word that can't be [precisely] translated. . . . Sisu refers not to the courage of optimism, but to a concept of life that says, 'I may not win, but I will give up my life gladly for what I believe.' . . . Sisu is the only word for the Finns' strongest national characteristic."

Niemi designed the airplane to allow people to build it in home workshops, but the first flight of the prototype Sisu 1 in 1958 was so successful that he decided not to sell plans or kits to homebuilders. Instead, Niemi modified the design for production as a finished, ready-to-fly sailplane. He changed the designation to Sisu 1A, lightened the wing structure, increased the size of the air brakes and added vents to them, and lengthened and widened the cockpit to hold bigger pilots. He also slightly increased the area of the V-tail and the deflection range of the rudder. He introduced the "plate-stringer" wing structure on the inboard wing sections to reduce weight without compromising the rigidity of the outer skin to maintain smooth surface contours that would facilitate laminar airflow.

Niemi paid careful attention to eliminating parasite drag when he conceived the Sisu. He opted for a V-tail, retractable tow hook and main wheel, and swept-forward wings shaped to a 65_3-418 laminar-flow airfoil. Rather than build washout, or twist, into the wingtips to reduce their tendency to stall, he leaned them ahead of the inner wing sections to insure that the ailerons, hinged just inboard of each wingtip, continued to bite the air and provide the pilot control during a stall. Washout—a slightly twisted wing—is a much more common technique, but Niemi eschewed it since it would have generated unacceptable drag. The wing surfaces had to be extremely smooth to maintain laminar flow over a broad speed range. He attached aluminum plates onto stringers and covered this assembly with relatively thick aluminum wing skins to prevent even the slightest buckle. Before applying finish paint, craftspersons checked the wing closely for minute imperfections, covered any depressions with filler paste, and sanded the patches smooth.

This obsession with a smooth finish enthralled one man who flew the Sisu to remark that at "first glance one may wonder and doubt the claimed performance figures. The tiny wings, the delicate fuselage keeping doubt alive, till one moves close and runs fingers over the surfaces of wings and fuselage, realizing at once that a perfectionist was at work." Pilots rated the cockpit roomy and comfortable, very important attributes in high-performance sailplanes that are easily capable of flying nonstop from early morning to dusk. They also praised the visibility directly in front of the pilot as outstanding.

To manufacture the Sisu, Nemi set up the Arlington Aircraft Company in the city of the same name about halfway between Dallas and Fort Worth, Texas. Construction began on the first four sailplanes in 1960. Pilots quickly snapped up the finished aircraft, but production costs surpassed profits, and Niemi had to sell the project to Philip J. Baugh, a retired air force pilot and soaring enthusiast from Charlotte, North Carolina. Baugh and Niemi moved production to Greenville, South Carolina, in 1963 and the firm became the Sisu Division of the Astro Corporation. Clifton McClure III managed Astro, and he and Baugh wisely decided to keep Niemi on as project engineer.

McClure was governor of the South Carolina division of the Soaring Society of America and an enthusiastic promoter of the sport of soaring sailplanes. He intended to carry through with Niemi's original plan for the Sisu and obtain Federal Aviation Administration type certification for the design, but the daunting costs of such work forced him to abandon the effort. Baugh generously underwrote production of six more Sisu aircraft, but profits never covered expenditures. Metalsmiths and technicians finished the tenth and last Sisu 1A in 1965.

Niemi took the second Sisu built (now at the NASM) into the air for the first flight on May 1, 1963. Less than two weeks later, on May 13, John J. Randall of Coral Gables, Florida, bought the sailplane. After several flights, Randall loaned the airplane to Dr. Joseph J. Cornish III at Mississippi State University in Starkville. Cornish teamed with test pilot Sean Roberts, and the two engineers conducted flight tests and measured and analyzed the Sisu's flight performance at a gross weight of 730 pounds. Their research yielded these numbers:

- Minimum sinking speed: 0.7 m (2.2 ft.) per second at 88 km/h (55 mph)
- Maximum glide ratio: 37:1 at 92 km/h (57 mph)
- Stall speed, flaps up: 79 km/h (49 mph)
- Stall speed, flaps 20 degrees down: 60 km/h (37 mph)
- Maximum airframe load factors: +6 G, -4.6 G
- Maximum Design Airspeeds:
    – Never-exceed: 277 km/h (172 mph)
    – Rough-air cruise: 192 km/h (119 mph)
    – Maneuver: 192 km/h (119 mph)
    – Flaps-down and auto/winch tow: 151 km/h (94 mph)

Above: Sisu designer Len Niemi stands beside Al Parker's 1A at Harris Hill, New York, in July 1967.

Below: Al Parker poses next to the Sisu before his ground crew tows the sailplane off the runway.

Cornish and Roberts believed that additional work on the sailplane could boost performance even higher. The men further refined the surface contours on this Sisu, but specific details about the techniques they used remain hidden.

On July 21, 1963, Alvin H. Parker bought the Sisu from Randall for $9,700. Parker was forty-five years old and a successful rancher and financier. He was also active in the oil business and operated a soaring flight school and a Schweizer sailplane dealership in Odessa, Texas. Parker had spent his entire life in West Texas except for a stint in the US Army during World War II, when he saw combat while commanding a Sherman tank in the First Armored Division fighting in North Africa and Italy. Parker had flown about five thousand hours in several different aircraft by that time and was ready to attempt the record soaring flight.

A year and ten days after he bought the Sisu 1A, on July 31, 1964, Parker took off from Ector County Airport north of Odessa and released from the tow plane just before 10:00 a.m. He found weak lift over the airport and glided north. "I gradually worked north along the Andrews Highway," he recalled. "Then to the west of the town of Andrews and up to about 2,800 feet northwest of the town. From there, I could look down on the family ranch land where I spent so many hours as a cow-puncher during my youth. If someone had told me that I would someday pilot a motor-less aircraft across this area, I would have told him he was 'loco.'"

Al Parker points a wingtip at the sky as he sails his Sisu 1A above the flat and arid scrub of west Texas.

By 2:00 p.m., Parker had descended to a few hundred feet above ground over northeastern New Mexico, still spiraling in search of lift but now at eye level with the canyon walls that cradle the sandy banks of the Cimarron River. "The quiet, desperate scratching techniques [to find lift] of the recent National Contest at McCook [in Nebraska, site of the 1964 national soaring contest held a few weeks previously] enabled me to get up and away [from the canyon]," Parker stated. "Placing thirty-eighth in a National Contest [at McCook] has its advantages—sometimes." Parker was referring to his poor showing at that meet and the lessons he had learned. He had finished nearly last out of forty pilots, but the experience taught him much about the hard work required to fly the Sisu in the weakest lift. On several occasions during the record flight, he put those lessons to good use.

Leaving the canyon far below, Parker soared the Sisu into Colorado at 8,000 feet above the Purgatoire River. He crossed the Arkansas River and passed the town of Hugo, cruising comfortably at 70 to 80 miles per hour. He chased but could not catch a dust devil madly twirling near Last Chance, Colorado, and then found the strongest lift of the day at the edge of a thunderstorm that carried him up at 1,000 feet per minute in muddy rain. He skirted the storm's edge in near-darkness and clouds and flew on instruments toward the goal he had declared before takeoff: Julesburg, Colorado, now just 65 miles away. Breaking into clear air revealed another strong storm between Parker and Julesburg. This menacing obstacle, combined with the growing darkness, forced him to abandon the declared goal and search northwest for the nearest airport. Parker spotted the welcoming blink of a rotating beacon but approached with care. If he miscalculated this final glide, he would have to attempt the only trick more difficult for a sailplane pilot than an off-airfield landing in unknown terrain: doing so at night.

Parker cruised the Sisu at 85 miles per hour for about fifteen minutes and reached the airport with a few hundred feet of altitude remaining. He touched down at 8:19 p.m., rolled to a stop, "struggled out of the Sisu after those ten and a half hours [of flight] and looked around for landing witnesses. There was not one in sight—just the rotating beacon, the runway lights, and I." A few minutes passed before two men appeared and told Parker that he had just landed at Kimball, Nebraska, 1,041.94 kilometers (647.17 miles) from Odessa.

The quiet finish to the spectacular flight and world distance record did not prepare Parker for the accolades that poured in. *Soaring* magazine called it the "Soaring Story of the Year—U.S. First to Break 1000 Kilometers." Parker affirmed that "the performance of the Sisu 1A was astonishing, actually beyond imagination. It would gain back, in one 360-degree chandelle, the altitude lost [by flying] at 120 mph between cu's [cumulus clouds] about 20 miles apart."

Two years later, in August 1966, a flight instructor qualified Al Parker's son Stephen, then fourteen years old, to fly the Sisu. Stephen made thirty-two flights, including a 281-mile flight from Odessa to Wheeler, Texas. On August 3, he flew 345 miles from Odessa to Farley, New Mexico. This flight earned him the coveted Diamond C soaring badge for distance, altitude gain, and endurance flying. Stephen was then the youngest American to earn the coveted badge. The two Parker men finished 194 flights in their Sisu, including 10 notable cross-country flights that spanned an average of 404 miles.

Al Parker sold the Sisu to Philip J. Baugh of Astro in August 1967 and then purchased another for his own use so that Baugh could donate the record 1,000-kilometer ship to the Smithsonian Institution, a deal the men had struck during the previous winter. On July 12, 1967, at a ceremony held atop Harris Hill near Elmira, New York, the birthplace of American competitive soaring, Baugh formally presented the Sisu 1A, serial number 102 and registered N1100Z, to the Smithsonian. The airplane was still equipped with the same instruments, radio, oxygen system, and other equipment that it had carried during the 1,000-kilometer flight in 1964.

The end of the 1960s also marked an end to the Sisu's dominant position in world competition glider flying. "As one decade ends and another begins," quipped *Soaring* magazine, "the all-metal Sisu finds itself fighting on against mounting odds—a tin soldier in an increasingly hostile world of fiberglass, where the prevailing language is German . . ." With sustained and ample support from the West German government, aerodynamicists, engineers, and enthusiasts at major universities designed and built motor-less aircraft made with fiberglass. By 1970, pilots flying the new sailplanes had surpassed those flying the Sisu and claimed the lead in competition soaring. Despite this dominance, no record was set or contest won that was quite as symbolic as Al Parker's 1964 flight.

Al Parker spent 10½ hours in this cockpit soaring 1,041.94 kilometers (647.17 miles) from Odessa, Texas, to Kimball, Nebraska, on July 31, 1964.

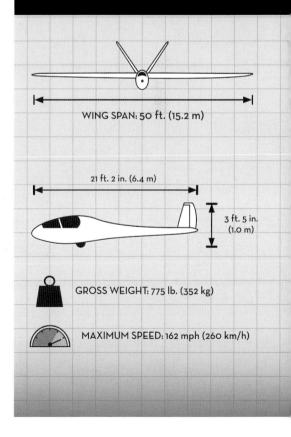

SPECIFICATIONS
ARLINGTON SISU 1A

WING SPAN: 50 ft. (15.2 m)

21 ft. 2 in. (6.4 m)

3 ft. 5 in. (1.0 m)

GROSS WEIGHT: 775 lb. (352 kg)

MAXIMUM SPEED: 162 mph (260 km/h)

BELL UH-1H IROQUOIS "HUEY" *SMOKEY III*

CHAPTER 23

BY ROGER D. CONNOR

Known almost universally as the "Huey," Bell Helicopter's Models 204 and 205 became iconic symbols of Cold War military actions. Used by all US military services, many allies, and numerous civilian operators, the Huey marked the milestone transformation of the helicopter from a niche tool to a primary instrument of Cold War doctrine. Its use in Southeast Asia during the Vietnam War made it one of the most enduring symbols of the conflict for supporters and opponents of American containment policy directed at the Soviet Union and its satellites. Its association with offensive combat actions as a transport for air assault troops and as a gunship belies its origins as an aerial ambulance and the saving of what might conservatively be estimated as one million lives during the decade of active American combat operations in Vietnam.

Today, the Huey soldiers on in heavily modified form with the US Marine Corps and many allies and is likely to continue to do so until the middle of the twenty-first century. It also supports nonmilitary public-service operations, such as law enforcement and firefighting—and industrial applications, such as logging and transport of oil-field workers, have made it nearly as successful in the commercial sector as it has been in the military sphere.

UH-1 Huey 65-10126 *Smokey III* on display at the Smithsonian National Air and Space Museum's Udvar-Hazy Center. With more than 2,500 combat hours in Vietnam, it is an exemplar for the experience of heliborne warfare in Southeast Asia.

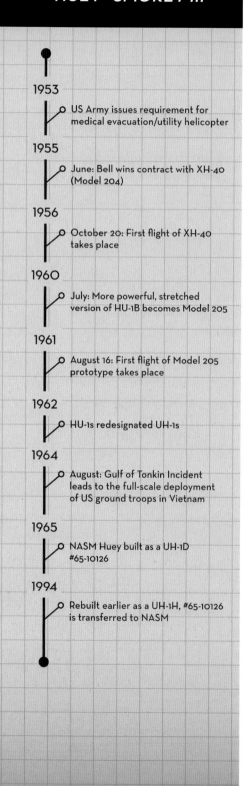

1953

US Army issues requirement for medical evacuation/utility helicopter

1955

June: Bell wins contract with XH-40 (Model 204)

1956

October 20: First flight of XH-40 takes place

1960

July: More powerful, stretched version of HU-1B becomes Model 205

1961

August 16: First flight of Model 205 prototype takes place

1962

HU-1s redesignated UH-1s

1964

August: Gulf of Tonkin Incident leads to the full-scale deployment of US ground troops in Vietnam

1965

NASM Huey built as a UH-1D #65-10126

1994

Rebuilt earlier as a UH-1H, #65-10126 is transferred to NASM

The Vietnam War remains the defining moment for the helicopter, which entered service in the late 1950s as the HU-1. This designation quickly gave rise to the "Huey" nickname in place of the army's formal—if scarcely used—official identification, Iroquois. The lack of conventional road networks and other infrastructure in South Vietnam meant that, as American involvement began in earnest in 1962, the helicopter took center stage in counterinsurgency operations against the Viet Cong guerillas who controlled much of the countryside. As the war progressed, military operations took the form of quick strikes into the countryside, much of which was covered by the marshes and rice paddies of the delta, dense interior jungle, or rugged hills and mountains. In this environment, helicopters—and the Huey in particular—stood out as the centerpiece of the army infantryman's experience in Vietnam. He might tremble with fear and anticipation as a Huey transported him into a contested landing zone; sigh with relief as one extracted him at the conclusion of an operation or delivered critical food, water, and ammunition; pray with thankfulness when one evacuated him after he was wounded; or even whoop with joy as a Huey gunship drove off a Viet Cong assault with its guns or rockets.

The Huey was not the only helicopter type used in Vietnam, but it was the most prominent. Around twelve thousand American helicopters served in Southeast Asia between 1961 and 1975, with Hueys making up slightly more than seven thousand of the total, the vast majority of which were employed by the US Army (followed distantly by the South Vietnamese and Australian armed forces, along with the US Marine Corps, navy, air force, and Central Intelligence Agency, in descending order). Hueys also made up 65 percent of the helicopter losses in Southeast Asia, with more than 3,300 shot down or destroyed by accidents in nearly equal quantities. Additionally, army and Corps commands in Vietnam deployed more than seven hundred AH-1G/J Cobra gunships, which were based on the Huey's dynamic components. No American military aircraft since World War II's B-24 bomber has seen greater numbers of production than the Huey.

What made the Huey so successful? Helicopter technology was rapidly maturing at a time when the US Army was looking to the helicopter as a way of keeping itself relevant in the Atomic Age, as well as establishing an air arm independent of the air force. While the helicopter made its combat debut in World War II, during which a handful saved between one hundred and two hundred soldiers on the battlefield, in the Korean War, medical evacuation helicopters were credited with carrying out forty thousand evacuations and helping to lower the fatality rate among wounded from 4.5 percent to 2.5 percent. Unfortunately, the reciprocating-powered Bell Model 47, the standard medical evacuation helicopter for the army and Marine Corps in Korea, could accommodate only two stretchers, both mounted externally. Not only were wounded exposed to the elements and subject to a much greater risk of hypothermia, medics could not attend to them in flight.

By November 1953, the army had developed requirements for a new medical evacuation helicopter that could also fulfill a general utility role. With the potential of being the largest helicopter contract awarded to date, fierce competition broke out among American manufacturers to meet the new specification. In June 1955, Bell Aircraft Corporation won the contract. Central to the army's new XH-40 design, which Bell called the Model 204, was the use of a jet turbine mounted above the passenger cabin.

On paper, the XH-40's groundbreaking Lycoming XT53 turboshaft (a jet turbine driving a gearbox) did not even fully match the performance of the existing reciprocating powerplants that had powered the army's helicopters to this point. However, the turbine configuration allowed a far more compact airframe so that the size of the aircraft was greatly reduced over earlier transport helicopters. The T53 also proved to have outstanding growth potential. While the XH-40 was intended principally to replace the diminutive Bell Model 47, it also nearly matched the capabilities of the Sikorsky H-19 that had served as the army's transport helicopter in the Korean War. Comparisons of performance metrics between the XH-40 and the H-19 did not reveal any significant advantages for the new design, but the Model 204 proved far easier to upgrade and modify—so much so that its evolved form in use at the height of the Vietnam War (the UH-1H) had 80 percent more payload than its Korean War predecessor with only a 32 percent increase in gross weight.

In addition, Bell's proprietary rotor system on the Huey was the simplest, lightest, most rugged, and most easily maintained in the industry. While this "semirigid" or "teetering" rotor system with the trademark Bell stabilizer bar imposed some maneuvering limitations not present in Sikorsky and other articulated rotor systems, it gave a clear operational advantage. Its wide chord and massive twin-blade rotor vortices also gave it the distinctive *whomp-whomp* auditory signature that became the enduring soundtrack of the Vietnam War.

When the XH-40 made its first flight on October 20, 1956, the army was in the midst of an identity crisis. With the ascension of the Eisenhower administration and John Foster Dulles's "New Look" doctrine, the Cold War faceoff with the Soviet Union became a matter of racing to possess strategic nuclear weapon systems—something that was out of the army's realm. The air force and, to a lesser extent, the navy were now the gatekeepers of American nuclear policy. The army and Marine Corps, desperate to retain their relevance on the hypothetical nuclear battlefield, began looking for new doctrines to justify their place at the budgetary table. In the decade after World War II, both services moved to embrace the helicopter as a way of ensuring mobility and survival in the Atomic Age.

The army's new policy originated from war hero and US Army Chief of Staff Maj. Gen. James Gavin. He unveiled his doctrine in an unlikely venue: the April 1954 issue of *Harper's Magazine*, in an article entitled "Cavalry, And I Don't Mean Horses." Gavin's "sky cavalry" embraced "airmobile" forces that would bypass enemy strongpoints by leapfrogging them with helicopters. Employing the post–Civil War ethos of the Great Plains cavalrymen fighting in the Indian Wars, Gavin foresaw the need for helicopter scouts and armed escorts as well as transports. As the H-40, soon redesignated with its HU-1 appellation, entered flight testing and its qualities became known, it became increasingly central to the implementation of this new doctrine. Gavin's chief architect was Maj. Gen. Hamilton Howze, who helped direct a series of boards in the five years leading up to direct American involvement in Vietnam that would establish the Huey as the centerpiece of the sky cavalry.

Though nuclear battlefields in Western Europe were Howze's primary concern, the sky-cavalry doctrine proved eminently

Opposite: The Bell Model 30 (front) and Model 47 (middle and rear) were the high points of Bell's early helicopter developments and established the pattern for the Huey's elegant mechanical simplicity.

Below: During the Vietnam War, helicopter production reached its peak at Bell's Fort Worth, Texas, Huey factory, which produced more than seven thousand before 1973.

suitable to limited warfare, and Southeast Asia in particular. This was a lesson the French had learned thoroughly with their experience of colonial wars in Indochina and Algeria. Their exposure on the limited road networks and trails of what would become Vietnam demonstrated the folly of overland logistics in the face of limited infrastructure, dense foliage, and a determined insurgency. The subsequent uprising in Algeria gave the French military an opportunity to show they had taken the implications of their defeat at the hands of the Viet Minh to heart by implementing new counterinsurgency strategies based on the helicopter that Howze and his boards were only just beginning to consider in a theoretical sense. This experience did not pass unnoticed in the United States, and as President Kennedy began calling on the army to support his policies by aiding the government of South Vietnam, it placed its helicopters and air mobility doctrine at the center of a strategy to defeat the Viet Cong.

The first helicopters to arrive in South Vietnam were reciprocating-powered models that predated the turbine era ushered in by the Huey. This was largely because of the Huey's and T53's to-be-expected teething troubles and the still-nascent nature of the doctrine used to support them. The fact was that the Model 204 was a good air ambulance and ultimately made a more than adequate gunship, but it was too small for a transport. Anticipating this even before Vietnam became a concern, Howze had pushed a stretched version that Bell implemented as the Model 205 and that entered service with the army as the UH-1D. It first flew several months before the first army helicopters arrived in Vietnam.

UH-1Ds were soon desperately needed as the reciprocating Vertol (formerly Piasecki) H-21s and Bell H-13s quickly showed their limitations in the devastating heat and humidity of Vietnam. The now-matured T53 had a great advantage in this climate. The Huey was the rising star of operations in Southeast Asia by the time of the Gulf of Tonkin incident in August 1964, and the Huey-centric sky cavalry was in place for its first major test during the Battle of Ia Drang in November 1965. Though costly in men and helicopters, the operation proved that even if small American units were cornered by large Viet Cong or North Vietnamese formations, they could be quickly reinforced at great loss to the enemy. Subsequent US strategy followed this model of baiting the enemy into a significant concentration and then enveloping them with helicopter-borne forces and devastating them with superior artillery and air support. The Huey would remain the centerpiece of these operations through the war.

Bottom left: The ability to operate almost anywhere in Vietnam gave American forces a decisive edge, but focused US policy on a seemingly endless cycle of indecisive "seek and destroy" campaigns.

Bottom right: Huey #65-10126 as it appeared in 1968, as *Smokey III* with the 11th Combat Aviation Battalion. Note the rubber bladder under the rear seat, which held the oil mixture to create smoke, and the ring around the engine exhaust, which created the smokescreen.

The National Air and Space Museum's Huey (65-10126) is an exemplar of the type's utility in Vietnam. It began as a UH-1D in 1965 and accumulated more than four years and 2,500 combat flight hours in theater. Initially attached to the 1st Cavalry Division, which had been the standard bearer for the sky cavalry's debut in Southeast Asia, it likely replaced an aircraft lost at Ia Drang. After serving as a "slick"—a Model 205 transport that lacked the external lethal accoutrements (guns and rockets) of the Model 204 gunship variants—this Huey joined the headquarters company of the 11th Combat Aviation Battalion, where it became a "smoker." Indicative of the wide utility of the Huey, smoke ships laid down smokescreens by using paraffin oil pumped from a large rubber bladder mounted in the rear cabin to a ring mounted on the exhaust stack that heated the oil into a thick cloud.

Nicknamed *Smokey III* by its crews, the museum's aircraft preceded air assault formations of regular slicks to screen them visually from potential ambushes during their initial insertion into a landing zone. As the formation of slicks returned for the extraction of ground forces, which might range in size from a small patrol of a dozen or so to a battalion consisting of many hundreds, the smoke ship led the way, flying at treetop level or lower and at agonizingly slow speeds to ensure optimal dispersion of the smokescreen. The nature of this mission naturally attracted enemy fire at an unusually high rate, with *Smokey III* and the small numbers of other battalion smokers taking more than their share of bullet holes.

An example of its operations is described in the text for the award of air medals to Specialists James Palmer and Robert Stidd (crew chief and gunner on *Smokey III*) for an action on February 13, 1968:

> For heroism while engaged aerial flight in connection with military operations against a hostile force: These men distinguished themselves while serving as door gunners on a smoke-ship. Committed to an extraction mission, the smoke ship was subjected to intense small arms, automatic weapons, and rocket propelled grenade fire. The smoke ship was responsible for laying a dense smoke screen between the fortified Viet Cong bunkers and the pick-up zone. Extremely vulnerable to heavy fire, the men delivered a high volume of accurate suppressive fire on each pass along the pick-up zone, thus allowing the infantrymen to load their wounded on the waiting slicks. Friendly casualties would have been tremendous if it had not been for the outstanding display of valor by these men. Their actions were in keeping with the highest traditions of the military service and reflect great credit upon themselves, their unit, and the United States Army.

The intensity of this aircraft's service is also related in the unit history of the 128th Assault Helicopter Company, recorded little more than a year later:

> On 1 May, while operating east of Cu Chi in support of Capital Military Assistance Command, "Smokey," the Tomahawk smoke ship experienced an engine explosion in flight. The aircraft was fully loaded with 500 lbs. of smoke solution and seven passengers. The Aircraft commander, Warrant Officer 1 Roland C. Lavallee, successfully auto-rotated without further damage to the aircraft. WO1 Levallee was commended with a "job well done" from several higher commands.

The engine explosion must have been severe, as the aircraft returned to the United States for repairs, though it returned to South Vietnam in the fall of 1969 for another six months of active service with the 118th Assault Helicopter Company as a conventional slick. Like many soldiers, 65-10126 returned to the United States in mid-1970 during President Nixon's drawdown of the US commitment in Southeast Asia. The aircraft continued in US National Guard service until its retirement in 1994.

SPECIFICATIONS
BELL UH-1H IROQUOIS "HUEY"
SMOKEY III

ROTOR DIAMETER: 48 ft. 3 in. (14.7 m)

41 ft. 5 in. (12.6 m)

13 ft. 7 in. (4.2 m)

EMPTY WEIGHT: 5,687 lb. (2,580 kg)
GROSS WEIGHT: 9,500 lb. (4,309 kg)

ENGINES: 1 x Lycoming T53, 1,250 shp
MAXIMUM SPEED: 127 mph (204 km/h)

LOCKHEED SR-71A BLACKBIRD

CHAPTER 24

BY F. ROBERT VAN DER LINDEN AND DIK DASO

For almost fifty years, the United States and the Soviet Union fought for the primacy of their political and economic systems. While occasionally, as in Korea and Vietnam, the struggle erupted in actual combat, the conflict mostly was fought during the tenuous peace of the Cold War. Airpower remained a crucial part of this battle. While not actively engaged in combat, aircraft—and, later, spacecraft—were instrumental in gathering crucial intelligence that allowed America's political and military leaders to make critical decisions necessary to national defense and protecting the peace.

During the early years of the Cold War, the subsonic Lockheed U-2 reconnaissance aircraft was the primary tool for gathering information on the USSR. When intelligence agencies were stymied in their efforts to ascertain the Soviet Union's intentions concerning the production of intercontinental ballistic missiles, U-2 missions photographed Soviet installations and helped analysts determine that the feared "missile gap" between the two nations did not exist. These flights were instrumental in guiding foreign policy.

Unfortunately, the downing of Francis Gary Powers and his U-2 by a Soviet SA-2 missile on May Day 1960 forced the cessation of these critical flights in the face of international condemnation and President Dwight Eisenhower's political

The SR-71 was designed to fly deep into hostile territory, avoiding interception with its tremendous speed and high altitude.

1959

Project OXCART approved for Mach 3 reconnaissance aircraft

1960

January 30: Lockheed wins contract for A-11

May 1: U-2 pilot Francis Gary Powers shot down over Soviet Union

1962

April 30: First flight of improved A-12

1964

February 24: President Johnson reveals existence of A-12

December 22: First flight of larger, two-seat SR-71

1966

SR-71 enters service

1967

May 3: First A-12 mission, over North Vietnam

1973

SR-71s provide intelligence on Yom Kippur War in Middle East

1974

SR-71s begin temporary operations in Europe

1982

Two SR-71s based in Great Britain

1990

January: SR-71s cease operations

March 6: SR-71 #972 sets transcontinental speed record and is retired to Smithsonian

1995

SR-71 program briefly revived

1999

SR-71 program ends

embarrassment. While a technological masterpiece, the U-2 was clearly vulnerable to Soviet defenses, particularly the SA-2 air defense system, which was designed to destroy high-altitude strategic bombers. America's intelligence community desperately needed a new system that was impervious to interception and could overfly sensitive sites. Not for the first time, Lockheed's famous "Skunk Works," under the brilliant leadership of Clarence "Kelly" Johnson, answered the call.

In the late 1950s, a secret committee was formed to examine the feasibility of building a reconnaissance aircraft capable of flying faster than three times the speed of sound at altitudes well above 80,000 feet and over a range of 3,000 miles. Under the chairmanship of E. M. Land of the Polaroid Corporation, the panel worked closely with Lockheed and Convair engineers, along with representatives of the air force and navy, to conclude that such an aircraft was indeed possible. With the support of the Central Intelligence Agency, the committee's recommendation to proceed was approved by President Eisenhower in mid-1959. Later that summer, Lockheed was selected to build this unique aircraft under the codename Project Oxcart. On January 30, 1960, Lockheed received approval to start construction on twelve of the new A-11 aircraft.

The engineering challenges were unprecedented. Cruising at Mach 3 would generate tremendous heat in excess of 550 degrees Fahrenheit, causing the failure of standard aircraft aluminum alloys. After thorough testing, a special titanium alloy was selected for its light weight, great strength, and heat resistance. It was also very expensive and very difficult

to work with, forcing Lockheed to create special drill bits and to build each aircraft virtually by hand.

In addition, special JP-7 fuel was developed to withstand the heat of 350 degrees Fahrenheit expected in the fuel tanks during flight, with nitrogen gas used to render any fumes inert. The fuel tanks were designed to expand during flight, which also caused leakage when they cooled down afterward and during refueling. Lubricants used in the aircraft were designed to operate from -40 to 600 degrees Fahrenheit (-40 to 316 degrees Celsius). In order to minimize the aircraft's radar signature, the A-11 was designed with special radar-absorbing laminated plastics, especially in the vertical stabilizers and the engine inlets. With so many changes, the aircraft was redesignated as the A-12.

As designed, the Lockheed A-12 was a sleek, twin-engined, delta-winged aircraft with unique "chines" running from the leading edge of the wings along the fuselage to the nose. Skunk Works engineers also optimized the A-12 cross-section design to exhibit a low radar profile, which Lockheed hoped to achieve by carefully shaping the airframe to reflect as little transmitted radar energy (radio waves) as possible and by applying special paint designed to absorb rather than reflect those waves. This treatment was one of the first applications of stealth technology, but it never completely met the design goals.

The carefully selected pilot sat in a pressurized cockpit and wore a full pressure suit at all times in case of an emergency at extreme altitudes. Two specially designed Pratt & Whitney J58 turbojets, each producing 34,000 pounds of thrust with afterburners, propelled this remarkable aircraft. These powerplants had to operate across a huge speed envelope in flight, from a takeoff speed of 207 miles per hour to more than 2,200 miles per hour. To prevent supersonic shock waves from moving inside the engine intake and causing flameouts, Johnson's team designed a complex air intake and bypass system for the engines. Because of production delays, however, the J58s were not ready in time for the A-12's first official flight on April 30, 1962 (although test pilot Louis Schalk had actually flown the single-seat A-12 several days earlier after he accidentally became airborne during high-speed taxi trials).

During these initial test flights, Pratt & Whitney J75s normally used on F-105s were fitted as an expedient. The A-12 first flew in earnest from a remote base at Groom Lake, Nevada, away from prying eyes. Select air traffic controllers were cleared concerning Project Oxcart so that the restricted airspace could be expanded and the flights unreported. Despite the military's best efforts, however, the A-12's existence was soon reported by airline pilots, the media, and the rumor mill. Eventually, President Lyndon Johnson revealed the aircraft publicly on February 24, 1964, referring to it as the A-11 in order to cover the existence of the classified A-12.

Even with its radical configuration and strict performance requirements, the A-12 proceeded through its testing with fewer difficulties than expected. Although several aircraft were lost during trials, the A-12 was declared operational in November 1965 and flew its first mission, an overflight of North Vietnam, on May 31, 1967. Operating from Kadena Air Force Base, Okinawa, the A-12 flew numerous more missions. On several occasions the North Vietnamese fired SA-2 missiles in response, but with no effect. A-12s were also used over North Korea during the USS *Pueblo* incident in January 1968.

Lockheed went on to build fifteen A-12s, including a two-seat trainer. Two were redesignated M-21s and modified to carry the D-21 reconnaissance drone mounted on a pylon between the rudders. The M-21 hauled the drone aloft and launched it at speeds

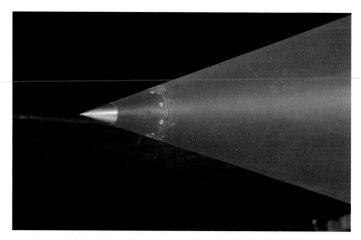

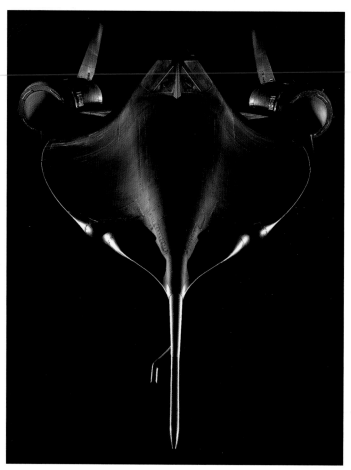

Top left: The pronounced spike over each engine's inlet was key to preventing the supersonic shockwave from entering the engines' compressors, thus allowing the SR-71 to travel at Mach 3+ speeds.

Bottom left: Lockheed's legendary Skunk Works got its name from a mysterious still in the *Li'l Abner* comic strip. The cartoon skunk has graced many Skunk Works aircraft since, including the National Air and Space Museum's SR-71.

Above right: The Blackbird had pronounced chines along the side of its nose to provide stability in yaw as well as extra room for fuel. The corrugated wings allowed for expansion as the airframe heated during supersonic flight.

high enough to ignite the drone's Marquardt ramjet motor. Although a great idea, the drones never worked properly.

Ironically, just as the A-12 was entering service, cost considerations forced the government to streamline its reconnaissance efforts. While the A-12 had been a civilian CIA project on which the military had worked closely, the US Air Force needed its own high-performance reconnaissance aircraft. Earlier, the air force had ordered an interceptor version of the A-12 to replace the defunct North American XF-108 Rapier project. Designated the YF-12, this new single-seat fighter carried four long-range AIM-47A air-to-air missiles intended to destroy high-altitude Soviet supersonic bombers.

When funding for the YF-12 ceased after the production of only four, the air force used the airframe of the A-12 to create a new reconnaissance aircraft: the SR-71. Slightly larger than the A-12, the SR-71 carried a two-man crew and first flew on December 22, 1964. It entered service in January 1966 at Beale Air Force Base, California, and quickly became the backbone of the air force's reconnaissance fleet. Although only thirty-two SR-71s were built, they gathered intelligence for more than thirty years over the world's hot spots, providing critical information that influenced many crucial decisions.

Because of extreme operational costs, military strategists decided that the more capable USAF SR-71s should replace the CIA's A-12s, which were retired in 1968 after only one year of operational missions, mostly over Southeast Asia. The air force's 1st Strategic

Reconnaissance Squadron (part of the 9th Strategic Reconnaissance Wing) took over the missions beginning in the spring of 1968.

After the air force began to operate the SR-71, the aircraft acquired the official name Blackbird for the special multipurpose paint that covered it. This was formulated to absorb radar signals, radiate some of the tremendous airframe heat generated by air friction, and camouflage the aircraft against the dark sky at high altitudes.

Experience gained from the A-12 program convinced the air force that flying the SR-71 safely required two crewmembers: a pilot and a reconnaissance systems officer (RSO). The RSO operated the SR-71's monitoring and defensive systems, including a sophisticated electronic countermeasures (ECM) system that could jam most acquisition and targeting radar. In addition to an array of advanced, high-resolution cameras, the aircraft could also carry equipment designed to record the strength, frequency, and wavelength of signals emitted by communications and sensor devices such as radar.

The SR-71 was designed to fly deep into hostile territory, avoiding interception with its tremendous speed and high altitude. It could operate safely at a maximum speed of Mach 3.3 at an altitude more than 16 miles, or 85,000 feet, above the Earth. To climb and cruise at supersonic speeds, the Blackbird's Pratt & Whitney J58 engines were designed to operate continuously in afterburner. While this would appear to dictate high fuel flows, the Blackbird actually achieved its best "gas mileage," in terms of air nautical miles per pound of fuel burned, during the Mach 3–plus cruise. That said, a typical Blackbird reconnaissance flight might require several aerial refueling operations from an airborne tanker. Each time the SR-71

This photograph provides a beautiful view of the "shock diamonds" coming from an SR-71's two powerful Pratt & Whitney J58 engines.

A parachute was deployed to help the SR-71 come to a stop upon landing after each mission.

refueled, the crew had to descend to the tanker's altitude, usually about 20,000 to 30,000 feet, thus slowing the airplane to subsonic speeds. As velocity decreased, so did frictional heat, causing the aircraft's skin panels to shrink considerably. Those covering the fuel tanks contracted so much that fuel leaked, forming a distinctive vapor trail as the tanker topped off the Blackbird. As soon as the tanks were filled, the jet's crew disconnected from the tanker, relit the afterburners, and again climbed to high altitude.

Air force pilots flew the SR-71 from Kadena Air Force Base, Japan, throughout its operational career, but other bases hosted Blackbird operations, too. The 9th Strategic Reconnaissance Wing occasionally deployed from Beale Air Force Base to other locations to carry out operational missions. Cuban missions were flown directly from Beale. The SR-71 did not begin to operate in Europe until 1974, and then only temporarily. In 1982, when the US Air Force based two aircraft at Royal Air Force Base Mildenhall to fly monitoring mission in Eastern Europe.

When the SR-71 became operational, orbiting reconnaissance satellites had already replaced manned aircraft to gather intelligence from sites deep within Soviet territory. Satellites could not cover every geopolitical hotspot, however, so the Blackbird remained a vital tool for global intelligence gathering. Blackbird crews provided important intelligence about the 1973 Yom Kippur War, the Israeli invasion of Lebanon and its aftermath, and pre- and post-strike imagery of the 1986 raid conducted by American air forces on Libya. In 1987,

SPECIFICATIONS
LOCKHEED SR-71A BLACKBIRD

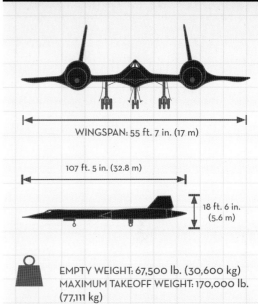

WINGSPAN: 55 ft. 7 in. (17 m)

107 ft. 5 in. (32.8 m)

18 ft. 6 in. (5.6 m)

EMPTY WEIGHT: 67,500 lb. (30,600 kg)
MAXIMUM TAKEOFF WEIGHT: 170,000 lb. (77,111 kg)

ENGINES: 2 x Pratt & Whitney J58-1 continuous-bleed afterburning turbojets, 34,000 lbf (151 kN) thrust each
MAXIMUM SPEED: 2,250 mph (3,620 km/h)

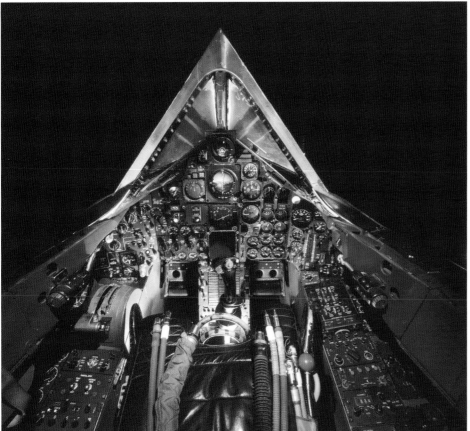

Above and left: The SR-71 cockpit was the epitome of analog instrumentation with its numerous dials and gauges.

Aerial refueling enabled the SR-71 to stay airborne for hours and to fly over thousands of miles of territory.

Kadena-based SR-71 crews flew a number of missions over the Persian Gulf, revealing Iranian Silkworm missile batteries that threatened commercial shipping and American escort vessels.

After a distinguished career, budget considerations forestalled further employment of the SR-71. As the performance of space-based surveillance systems grew, along with the effectiveness of ground-based air-defense networks, the air force began losing enthusiasm for the expensive program, and the 9th Strategic Reconnaissance Wing ceased SR-71 operations in January 1990. Despite protests from military leaders, Congress revived the program in 1995, but continued wrangling over operating budgets led to final termination. The National Aeronautics and Space Administration (NASA) retained two SR-71As and the one SR-71B for high-speed research projects, which they flew until 1999.

The Blackbird did not retire quietly, however. On March 6, 1990, SR-71 serial number 64-17972 made history when it set a transcontinental speed record, taking off from Palmdale, California, at 4:30 a.m. and heading over the Pacific Ocean, where it met a KC-135 and refueled. After a long delay to correct a faulty fuel gauge, the SR-71 climbed to its cruising altitude above 80,000 feet and accelerated to more than Mach 3. At 9:00 a.m. it crossed the West Coast, and one minute later officially entered Los Angeles airspace. Thirty-eight minutes after that the Blackbird passed Kansas City, and at 10:06:18 it reached Washington, D.C. Two minutes later it was over the Atlantic. After refueling again, the SR-71 returned to Washington Dulles International Airport. The crew of Lt. Cols. Joe Vida and Ed Yeilding made one pass over a crowd of VIPs and astonished airline passengers. Instead of landing immediately after the second approach, the crew lit the afterburners in a dramatic farewell gesture before landing and turning the aircraft over to the National Air and Space Museum. Among the dignitaries on hand to greet them was Ben Rich, the director of Lockheed's Skunk Works. On its last flight, the SR-71 flew coast to coast, covering 2,404.05 miles in sixty-seven minutes and fifty-four seconds at 2,124.05 miles per hour, and from Los Angeles to Washington, D.C., a distance of 2,299.67 miles, in sixty-four minutes and nineteen seconds at 2,144.83 miles per hour.

Today the SR-71 still holds the record as the world's fastest jet-powered aircraft.

After its retirement from the US Air Force, the SR-71 remained in service with NASA for supersonic testing.

AÉROSPATIALE-BAC CONCORDE

CHAPTER 25

BY F. ROBERT VAN DER LINDEN

It began with a dream—a dream of a new age in air travel in which the boundaries of time and distance would be shattered forever. The dream of supersonic passenger air travel was conceived in the 1950s, developed in the 1960s, and realized in the mid-1970s. And for twenty-seven years, the graceful Anglo-French Concorde carried world travelers across the Atlantic Ocean in great comfort at twice the speed of sound.

While the dream was real, it was so only for the world's privileged elite. The Concorde was not a machine accessible to the average citizen; high development and operating costs prevented it from providing supersonic flight to the wider public. But for a while, it looked promising. It looked like the future.

In the 1950s, jet propulsion revolutionized air travel. First the de Havilland Comet and later the Boeing 707 greatly increased the speed of travel from 350 to over 600 miles per hour. Airlines and customers flocked to the new jet airliners as travel times were cut dramatically and the seat-mile costs to the airlines dropped. The conclusion drawn by engineers, managers, and politicians seemed clear: the faster the better.

Air France gave the National Air and Space Museum their flagship Concorde, F-BVFA, which is now a centerpiece at the Udvar-Hazy Center.

TIMELINE

AÉROSPATIALE-BAC CONCORDE

1962

November 29: Great Britain and France agree to build supersonic transport

1969

March 2: First flight of Concorde

1976

January 21: Concorde enters service with Air France and British Airways

May 24: Special service to Washington, D.C., is flown

November 22: Service to New York begins

1989

Air France agrees to donate Concorde to Smithsonian

2000

July 25: Concorde suffers first and only fatal accident; fleet grounded

2001

September 11: Terrorist attacks in United States cause severe decline in air travel

November 7: Concorde returns to service

2003

April: Airbus withdraws technical support for Concorde

May 31: Last scheduled flight of Air France Concorde

June 12: Air France flies flagship Concorde F-BVFA to Smithsonian

October 24: Last scheduled flight of British Airways Concorde

December 15: Concorde F-BVFA goes on display at Udvar-Hazy Center

In Europe, enterprising designers in Great Britain and France were independently outlining their plans for a supersonic transport (SST). In November 1962, in a move reminiscent of the Entente Cordiale of 1904 (the beginnings of the French–British alliance), the two nations agreed to pool their resources and share the risks of developing and building this new aircraft. They also hoped to highlight Europe's growing economic unity and aerospace expertise in a dramatic and risky bid to supplant the United States as the leader in commercial aviation. The aircraft's name reflected the two nations' shared hope for success through cooperation—Concorde.

Quickly the designers at the British Aircraft Corporation and Sud Aviation, later reorganized as Aérospatiale, settled on a slim, graceful form featuring an ogival delta wing (that is, with a curved leading edge) that possessed excellent low- and high-speed handling characteristics. Power was to be provided by four massive Olympus turbofan engines built by Rolls-Royce and SNECMA. Realizing that this first-generation SST would cater to the wealthier passenger, the Concorde's designers created an aircraft that carried only one hundred seats in tight, four-across rows. They assumed that first-class passengers would flock to the Concorde to save valuable time while economy-class passengers would remain in larger, slower subsonic airliners. They also hoped that the Concorde would be only the first in a fully developed line of progressively larger SSTs that would carry more than one hundred passengers each and, thus, gradually reduce the cost of travel and greatly widen the market beyond the economic elite.

Despite mounting costs that constantly threatened the program, construction continued with exactly 50 percent of each aircraft occurring in each country. In 1969, the first Concorde was ready for flight. With famed French test pilot André Turcat at the controls, Concorde 001, which was assembled at Toulouse, took to the air on March 2, 1969. Although the Soviets had flown their version of the SST first, the Tupolev Tu-144 had been rushed into production and suffered from technological problems that would never be solved. Following the Concorde's successful first flight, more prototype and preproduction Concordes were built and thoroughly tested. By 1976, the first production Concordes were ready for service.

But all was not rosy. During this time America sought to produce its own bigger and faster SST. After a contentious political debate in 1971, however, the federal government refused to back the project, citing environmental problems, particularly noise, the sonic boom, and engine emissions that were thought to harm the upper atmosphere. Anti-SST political activity in the United States delayed the granting of landing rights, particularly in New York City, causing further delays.

More ominously for Concorde, no airlines placed orders. Despite initial enthusiasm, the airlines dropped their purchase options once they calculated the Concorde's operating costs. Consequently only Air France and British Airways—the national airlines of their respective countries at that time—flew the sixteen production aircraft, and only after acquiring them from their respective governments at virtually no cost.

Concorde service began in January 1976, and by November these stylish SSTs were flying to the United States. A technological masterpiece, each Concorde smoothly transitioned to supersonic flight with no discernable disturbance to the passenger. In service, the Concorde cruised at twice the speed of sound between 55,000 and 60,000 feet—so high that passengers could see the curvature of the Earth. The Concorde was so fast that, despite an outside temperature of less than -69 degrees Fahrenheit, the aircraft's aluminum skin

would heat up to more than 248 degrees Fahrenheit and the Concorde actually expanded 8 inches in length with the interior of the windows gradually growing quite warm to the touch. All the while, each passenger was carefully attended to as they enjoyed a magnificent meal. Interestingly, no movies were shown; by the time the aircraft got to cruising altitude and the meal service was completed, there was no time. Instead, the passengers watched the Mach meter on the bulkhead, cheering as the aircraft reached Mach 1 and then, the goal of every flight, Mach 2, twice the speed of sound. Transatlantic flight time was cut in half, with the average flight taking less than four hours.

The Concorde became the choice of millionaires, celebrities, and rock stars. Famed cellist Mstislav Rostropovich always reserved two seats: one for himself and the other for his cello. As part of the famous 1984 "Band Aid" concerts in London and Philadelphia that raised millions of dollars for African famine relief, rock star Phil Collins performed in London, hopped on a Concorde, and flew to Philadelphia in time to perform there.

British Airways and Air France were the only airlines to purchase the Concorde.

But eventually, harsh economic realities forced Air France and British Airways to cut back their already limited service. Routes from London and Paris to Washington, D.C., Rio de Janeiro, Caracas, Miami, Singapore, and other locations were cut, leaving only the transatlantic service to New York. And even on most of these flights, the Concorde was only half full, with many of the passengers flying as guests of the airlines or as upgrades. With the average roundtrip ticket costing more than $12,000, few could afford to fly this magnificent aircraft. Operating costs escalated as parts became more difficult to acquire, and, with an average of one ton of fuel consumed per seat, the already-small market for the Concorde gradually grew smaller.

Despite the excellence of the Concorde's design, its operators realized that its days were numbered. In 1989, in commemoration of the two hundredth anniversaries of the French Revolution and the ratification of the US Constitution, the French government sent a copy of the Declaration of the Rights of Man and of the Citizen to the United States. Appropriately, this famous document was delivered on the Concorde and with it a promise from Air France to give one of these aircraft to the people of the United States through its eventual inclusion in the collection of the Smithsonian Institution's National Air and Space Museum. It was assumed that that day would be well in the distant future. One accident changed that.

For most of its career, the Concorde was statistically the safest commercial airliner in the world, having had no fatal accidents in almost twenty-five years of service. Then on July 25, 2000, a main wheel of Air France Concorde F-BTSC struck a metal shard on the runway during takeoff at Charles de Gaulle Airport in Paris. Fragments from the exploding

The Soviet Union flew the first supersonic transport, the Tu-144, but technical problems prevented its regular use as an airliner.

tire struck a main fuel tank in the wing, causing it to release thousands of pounds of fuel. The hot engine exhaust ignited the kerosene instantly, causing the doomed airliner to crash nearby with the loss of all 109 people on board and 4 on the ground in the nearby village of Gonesse.

A similar incident had occurred several years before on a British Airways Concorde flying out of Washington Dulles. Fortunately, the tank in that case was empty, and no fire ensued; the aircraft was grounded pending an investigation into the cause.

British Airways was eager to repair and modify its Concordes, while Air France, more concerned about the aircraft's diminishing value and increasing costs, was less enthusiastic. Nevertheless, the Concorde returned to service amid much fanfare in the summer of 2001,

Boeing's Model 2707 was an attempt at a supersonic transport to compete with the Concorde. It failed when the US Congress refused to fund the program.

AÉROSPATIALE-BAC CONCORDE

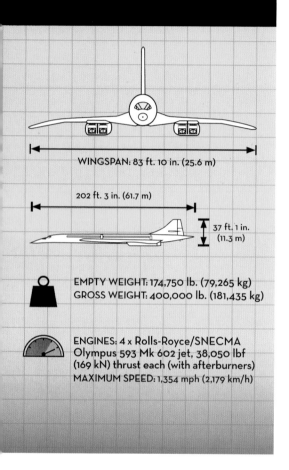

WINGSPAN: 83 ft. 10 in. (25.6 m)

202 ft. 3 in. (61.7 m)

37 ft. 1 in. (11.3 m)

EMPTY WEIGHT: 174,750 lb. (79,265 kg)
GROSS WEIGHT: 400,000 lb. (181,435 kg)

ENGINES: 4 x Rolls-Royce/SNECMA
Olympus 593 Mk 602 jet, 38,050 lbf
(169 kN) thrust each (with afterburners)
MAXIMUM SPEED: 1,354 mph (2,179 km/h)

The Concorde's tight cockpit was packed
with instrumentation.

only to face the economic disaster that struck the international airline community following the terrorist attacks of September 11 which caused a dramatic decrease in airline travel around the world.

Nevertheless, Concorde service continued until April 2003. It was then, with maintenance costs spiraling ever upward and the costs of new parts becoming prohibitive, that Airbus Industries, the successor to Sud Aviation, made the difficult but logical decision to ground the aircraft permanently. Air France and British Airways, the only airlines to own and operate the Concorde, reluctantly announced the cancellation of supersonic service. But this was not the end of the story.

Fourteen years after Air France promised to donate a Concorde to the National Air and Space Museum, that promise was fulfilled. In April 2003, Air France president

Jean Cyril Spinetta informed the museum that Concorde service would end on May 31. When Air France pledged a Concorde to the Smithsonian in 1989, no one at the museum expected it to arrive until 2007 at the earliest. Fortunately, Smithsonian leadership had approved the construction of four additional bays at the new Steven F. Udvar-Hazy Center then under construction at Washington Dulles International Airport, giving the museum the extra room needed to install the Concorde amid 170 plus other aircraft. Only the delivery of the aircraft remained.

On June 12, Air France delivered Concorde F-BVFA to Washington Dulles on the last supersonic flight for the airline. F-BVFA was the most distinguished Concorde in their fleet, having opened Air France SST service to New York, Washington, Caracas, and Rio de Janeiro. It was the pride of their fleet, and the NASM was fortunate to receive it.

Onboard that day were sixty passengers, including Gilles de Robien, the French Minister for Capital Works, Transport, Housing, Tourism, and Marine Affairs; Spinetta and several past Air France presidents; and former Concorde pilots and crew members. Mstislav Rostropovich was there too, but without his cello. While it was technically a ferry flight to the NASM's new Udvar-Hazy Center, Air France treated it no differently than its regular commercial Concorde service. In other words, the flight and the service were magnificent. On landing, in a dignified yet bittersweet ceremony, Spinetta signed over Concorde "Fox Alpha" to the museum for permanent safekeeping. The champagne flowed freely that afternoon.

The National Air and Space Museum is honored to have this beautiful aircraft in its collection at the Udvar-Hazy Center—not far from where it first landed with paying passengers back in 1976.

The Smithsonian National Air and Space Museum's Concorde lands at Washington's Dulles Airport on its last flight. The beautiful Concorde featured an ogival (curved leading edge) delta wing for excellent subsonic and supersonic flight. Its nose was lowered on landing for better pilot visibility.

GENERAL ATOMICS MQ-1L PREDATOR

CHAPTER 26

BY ROGER D. CONNOR

As an aerospace milestone, the Predator marked several significant transformations underway at the beginning of the twenty-first century. The first was the dramatic shift from manned aircraft to remotely piloted aircraft systems (RPAS). This shift had occurred slowly over much of the twentieth century as cruise missiles, target drones, and autopilots narrowed the role of onboard human pilots. The MQ-1 and other RPAS consist of "air vehicles" and ground-control equipment connected to the air vehicle by radio and satellite data links. In most RPAS, including the Predator, humans are essential to their routine operation. While no one flies on the Predator, and it often cruises under control of an autopilot, most of its functions occur at the hands of a pilot, sensor operator, and mission intelligence coordinator in the ground-control station. In this way, the Predator is more "manned" than many other combat aircraft.

The second milestone represented by Predator was the ability of its operators to fly it remotely and fire precision weapons in real time from ground stations on the opposite side of the planet. This encouraged development of new strategic doctrines that allowed persistent observation and aerial attack against individual targets with the same weapon system. Previously, the use of separate systems for aerial reconnaissance and for weapons delivery, often involving

When Predator 3034 went on display in 2008, the MQ-1 had transitioned from a military curiosity to the focus of a geopolitically contentious debate over the use of remotely controlled weapons.

1994

○ January: General Atomics awarded Predator development contract

○ July 3: First flight of Predator

1995

○ Summer: First operations of Predators over Balkans

2001

○ February 16: First successful firing of Hellfire missile from Predator

2001–2003

○ September 2001–January 2003: NASM's Predator #3034 flies 164 sorties over Afghanistan

2011

○ March 3: US Air Force takes delivery of last Predator

2013

○ October 22: Predators achieve 2 million flight hours of operations

different military organizations, had created operational and bureaucratic delays of hours, or often days, that allowed targets to escape attack or resulted in mistaken targeting. In military parlance, the Predator "closed the sensor-to-shooter cycle" (the time between the identification of the target and the strike) and "accelerated the kill chain." The use of the Predator as an aerial sniper ushered in a new form of warfare with its own cultural and geopolitical implications.

Third, the Predator's rise encouraged (but did not cause) a massive upsurge in the use of RPAS by the United States and created an unmanned arms race among other nations seeking similar capabilities. While the MQ-1 was not the most numerous military RPAS of the first decade and a half of the twenty-first century, the nature of its operations gave it a higher public profile than any other and colored the generic term *drone*. The dual nature of the Predator as an observation system and precision-strike aircraft complicated discourse surrounding other so-called drones as they came into greater civil and commercial use.

The terrorist attacks of September 11, 2001, ushered in a sustained period of limited war for the United States in Asia and Africa that catapulted the Predator to prominence. The ability of hostile nonstate organizations such as al-Qaeda, the Taliban, and the Islamic State to base in nations ostensibly allied with, or neutral to, the interests of the United States posed a significant strategic challenge during the War on Terror. Whereas airstrikes, cruise-missile strikes, and special-operations forces were the primary mode of American operations against terrorist organizations before 2001, they often engendered significant resistance and resentment in local populations and thus were not politically sustainable. Because Predators operated out of sight and were unmanned, and because their missiles caused less unintended death and destruction than other means, the MQ-1 and its further development, the MQ-9 Reaper, operated with less political resistance than traditional forms of intervention.

The resulting "drone campaigns" over Afghanistan, Pakistan, Yemen, Somalia, Iraq, Libya, and Syria have left a complex legacy, with significant claims of success in eliminating terrorist leaders but also considerable concern and resentment arising in much of the world over collateral damage resulting in the deaths of bystanders. There has also been a tendency to wrongly assign blame to Predators for strikes carried out by other means, such as by combat aircraft of the sovereign nation where the MQ-1 was used. For some, the Predator represents a cold-hearted disregard for civilians. However, for American policymakers, military commanders, and a majority of the American public, "drone strikes" are the most efficient and effective means of keeping deadly terrorists at bay—and do so with no risk to American service personnel.

The Predator had an unconventional and rapid development cycle unusual in modern American military aircraft. The MQ-1's origins go back to a garage project by Israeli emigrant Abraham Karem. By 1983, he had developed a small long-endurance tactical reconnaissance unmanned aerial vehicle (UAV) prototype called the Albatross for the Defense Advanced Research Projects Agency (DARPA). Five years later, further development resulted in a more advanced design, the Amber, with a superb loitering time due to its high-aspect-ratio,

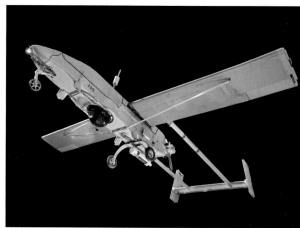

sailplane-like wings. A further development, the GNAT 750, resulted in a production-worthy design. Karem's company and the GNAT 750 were soon acquired by General Atomics.

The Central Intelligence Agency (CIA) flew the GNAT 750 in operations over Yugoslavia in 1993 and 1994. The program suffered from several technical issues, but it held enough promise that the Department of Defense expressed interest in a larger, more capable enhanced version of the GNAT for medium-altitude tactical reconnaissance, soon designated the RQ-1 Predator. By 1995, the army was operating it over Yugoslavia.

Initially, the navy-helmed UAV Joint Program Office had controlled Predator development, but the air force soon saw the Predator as an interim replacement for a shortfall in tactical reconnaissance aircraft with the added benefit of a live satellite video link. The air force gained developmental and operational control of the program from the other services in 1996. Authority for Predator development came under the 645th Aeronautical Systems Group, nicknamed "Big Safari," which had responsibility for rapid development of reconnaissance systems outside conventional "mil-spec" airworthiness standards, resulting in a capable but fragile aircraft. In the late 1990s, Big Safari expanded the Predator's capability to include a laser designator to illuminate targets and guide weapons dropped from other aircraft, reducing the sensor-to-shooter cycle. In 1999, this system had its first significant test during Operation Allied Force in Kosovo. One lesson learned was that the sensor-to-shooter cycle was still too great, as sometimes strike aircraft struggled to locate targets spotted by Predator. The solution was to explore options for arming the Predator with its own ordnance.

By 2000, concern over the rising threat of the al-Qaeda terrorist organization and its leader, Osama bin Laden, encouraged Big Safari to accelerate the schedule for arming the Predator with the AGM-114 Hellfire laser-guided missile, originally developed for antitank helicopters. Given the requirement to make executive-level decisions about priority terrorist targets that might appear only briefly, the ability to watch Predator video and control the aircraft from the other side of the world in real time took on increased urgency. While Big Safari continued development in the United States, it also secretly operated several Predators from a base in Uzbekistan with the CIA during the summer and fall of 2000 to locate Osama bin Laden in Afghanistan. This paid off in September, with two and possibly three successful sightings, but President Clinton's administration had not yet made a decision on direct military action against bin Laden, so the Predators soon returned to the United States. These operations also validated what would become the standard operational profile of the Predator in which a launch-and-recovery team at the operating airfield oversaw the takeoff and landing, but then passed control to a remote ground-control station via satellite, where a different crew managed the mission.

Predator number 3034 demonstrated the first static firing of a Hellfire on January 23, 2001. In February it began airborne trials of the Hellfire, and in April, with the addition of a new sensor ball and laser targeting system mounted in the nose, number 3034 had evolved into what would become known as the MQ-1—the armed Predator that would become the mainstay of antiterror UAV operations for more than a decade to come.

Above: With its upper cowling removed, a NASA-owned Predator B (MQ-9) showcases the Ku-band satellite dish that makes the Predator such a capable platform for real-time intelligence.

Opposite: Before General Atomics began producing remotely piloted aircraft systems (RPAS), Israeli designs like the RQ-2 Pioneer were the most capable types on the battlefield. This museum example was the first remotely piloted aircraft to have humans surrender to it.

Continuing concerns over the legality of targeting an individual like bin Laden with the armed Predator delayed further deployments to Uzbekistan through the summer of 2001. On September 11, just as number 3034 and two other MQ-1s were undergoing final trials before deployment, the worst-case al-Qaeda attacks occurred in New York and Washington, D.C.

By October 7, 2001, number 3034 was on nighttime patrol with two Hellfires near Kandahar, Afghanistan, then under control of the Taliban, the state patron of al-Qaeda in Afghanistan. On this sortie, number 3034's crew successfully located Taliban leader Mullah Mohammed Omar's convoy, maintained a firing position, and requested clearance to engage. In an incident that has been highly debated ever since, authorization from the US Central Command was delayed, and Omar entered a structure. Concerns about potential incidental causalities and damage led to a decision to strike a Taliban security vehicle nearby in a ploy to flush Omar out into the open, which failed. This was the first Hellfire strike from a Predator and the first successful operational use of a precision-guided, UAV-delivered weapon, though other unmanned aircraft had been employed in offensive operations. During World War II, UAVs loaded with explosives were used as weapons by both the United States and Germany. The US Navy employed the unmanned Interstate TDR-1 using free-fall bombs against Japanese targets. At the end of the Vietnam War, the Ryan Model 147 reconnaissance drone was tested extensively with laser-guided missiles and bombs, but never reached operational status.

Predator number 3034 flew 164 operational sorties over Afghanistan between September 2001 and January 2003, with missions lasting an average of over fourteen hours and some over twenty hours. Between August and November 2002, during the middle of its operations from Uzbekistan, number 3034 undertook a detached deployment to another operational site. This could have been either Operation Southern Focus over Iraq or deployment to Djibouti in the Horn of Africa to support a CIA surveillance program over Yemen, where it flew thirty-two missions. Given this aircraft's close association with the CIA and the Big Safari program and the timing of initial operations from Djibouti, the latter scenario is more likely. During its probable time there, a Predator Hellfire strike killed Qaed Salim Sinan al-Harethi, who had masterminded the attack on the destroyer USS *Cole* on October 12, 2000. The National Air and Space Museum acquired Predator number 3034 in 2004 on the basis of its pivotal role in introducing armed RPAS into combat.

Given the small number of strikes made by Predators compared to manned aircraft, the impact on doctrine was enormous. The success of the type in locating and killing top enemy leaders made it a favorite weapon of national security advisors and military commanders alike. In April 2001, the US military had only ninety nontarget drone UAVs in service, seventy-five of which were small battlefield observation types: the RQ-2 Pioneer and the RQ-7 Shadow. The other fifteen were Predators. By 2013, the number had tripled. Ten years after that, the US military had nearly eleven thousand UAVs on inventory. The Predator alone does not account for this profusion, but it unquestionably established the potential of the UAV to shape the battlefield and geopolitics in ways that no aircraft, manned or unmanned, had done before.

Previously, military strikes consisted of fast jet pilots arriving over a chaotic scene with little time to understand the situation, releasing heavy ordnance, and then quickly departing. Accuracy in such engagements could be problematic, particularly with terrorists

Opposite: An MQ-1 Predator armed with AGM-114 Hellfire missiles flies a combat mission over southern Afghanistan. The MQ-1 is providing interdiction and armed reconnaissance.

Below: Experience has shown that strict division of duties between pilots and sensor operators is essential to maintaining situational awareness in the current generation of remotely piloted aircraft.

or insurgents who blended with the local populace. Instead, with the ability to remain airborne for up to forty hours (though operational missions rarely go much beyond twenty), Predator pilots and sensor operators could understand the ground situation far more clearly than in any previous aerial platform. The Hellfire missile, while powerful, also has a narrow blast effect, which made possible precision strikes that were impossible from manned aircraft. At a typical operating altitude of 15,000 feet above the terrain, the Predator is silent and invisible to those on the ground (though not stealthy to radar).

This new mode of warfare came with new problems. Because the Predator bypassed the normal procurement process and did not have to meet conventional military standards for ruggedness and reliability it entered operations in something akin to a prototype phase of development. This contributed to an exceptionally high loss rate that reached twenty-eight per 100,000 flight hours (eventually dropping to less than a quarter of that). Another concern was the "soda straw" effect of viewing the world through the tight focus of the camera lens and thus missing important nearby activity. This pushed investments in multiple-camera-array sensor systems that could monitor larger areas and use computer algorithms to highlight likely areas of concern, such as a vehicle driving in a certain manner or in the appearance of weapon system.

USAF Predator production ended in 2011 with 268 airframes completed. Additional unarmed airframes were made available to allies, including the United Kingdom and Italy. The army began development of a refined derivative, the MQ-1C Gray Eagle, which began operations in 2012. Even by the time of Operation Allied Force in 1999, the air force was aware that it needed a more capable and refined version of the Predator, so General Atomics began work on the "Predator B," which entered operations in 2007 as the MQ-9 Reaper and slowly began replacing the Predator. The Reaper has 25 percent less endurance the MQ-1 but can carry considerably more ordnance: four Hellfire missiles and two 500-pound bombs.

The Reaper and Predator are well matched to the nature of the Global War on Terror. For the most part, they have operated against terrorists and insurgents who lack aircraft and air defenses. However, as operations in Iraq, Syria, and Kosovo have shown, they are extremely vulnerable when opposed by a capable enemy, as they are slow and cannot maneuver aggressively. In 2002, the air force even adapted a Predator to carry Stinger missiles and attempted an air-to-air engagement with an Iraqi MiG-25—a dogfight that resulted in the loss of the Predator.

The inability of Predator and Reaper to operate in contested airspace with effective enemy air defenses highlights the advances required for RPAS to maintain their operational significance. Jamming poses a significant threat to the Predator's data links and GPS navigation, so future systems require significant advances in artificial intelligence and inertial navigation as well as faster and stealthier airframes. Another challenge is cultural—who is considered a pilot? Initially, most RPAS pilots for the US Air Force were experienced combat pilots, but demand soon exceeded supply and the military services began training non-pilot operators. This has created organizational frictions in the military over who has the privileges of pilot status in a world where unmanned and autonomous operations are increasingly important. Regardless, the Predator and subsequent RPAS have dramatically changed the strategy and tactics of limited war in the twenty-first century.

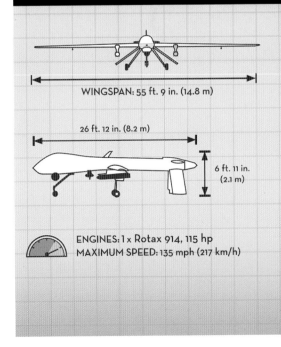

SPECIFICATIONS
GENERAL ATOMICS MQ-1L PREDATOR

WINGSPAN: 55 ft. 9 in. (14.8 m)

26 ft. 12 in. (8.2 m)

6 ft. 11 in. (2.1 m)

ENGINES: 1 x Rotax 914, 115 hp
MAXIMUM SPEED: 135 mph (217 km/h)

IMAGE CREDITS

INDEX

Abbott, Charles G., 51
Acosta, Bert, 29
Adams, Baxter, 27
Adams, Michael, 166–167
Aerial Experiment Association (AEA), 27
Aero Enterprises, 69
aerodynamic heating, 162–164, 182–183, 192–193
Aeronca C-2, 65
Aérospatiale, 192
Aérospatiale-BAC Concorde, 190–197
ailerons, 27
Air Commerce Act (1926), 130
Air France, 194–197
Airbus Industries, 196
al-Qaeda, 201–202
Alcock, John, 37
Allen, Eddie, 129
Allen, William, 151
American Airlines, 80
American Airman's Association (AAA), 130–131
Anderson, Orvil A., 71–74
Arado Ar 234, 110
Arlington Aircraft Company, 171
Arlington Sisu 1A sailplane, 168–173
Armstrong, Neil, 166
Arnold, Heinz, 115
Arnold, Henry "Hap," 74, 101–103, 122, 136
Arrowing A-2 Chummy, 64–65
Astro Corporation, 171
atomic bombs, 119, 122–124

B. F. Goodrich Company, 61
Baldwin, Thomas Scott, 27
balloons, 71–75
Barnes, Florence "Pancho," 55
Baugh, Philip J., 171, 173
Baumgartner Carl, Ann G., 107
Baumgartner, Felix, 74–75
Becker, John, 162
Bell Aircraft Corporation, 102–104, 137, 177
Bell, Alexander Graham, 27
Bell, Lawrence D., 102, 141
Bell Model 47 helicopter, 176
Bell UH-1 "Huey" *Smokey III*, 174–179
Bell XP-59A Airacomet, 100–107, 137
Bell XS-1/X-1, 134–141, 162, 164
Bellanca, Giuseppe, 46
Beser, Jacob, 125
Big Safari program, 201–202
Bixby, Harold, 48

Blair, Charles F., 98–99
Blanchard, Jean-Pierre, 72
Blériot, Louis, 23–25
Blériot Type XI monoplane, 23–27
Boeing 247-D, 78
Boeing 367-80 ("Dash 80"), 148–153
Boeing 707, 152–153
Boeing Aircraft Company, 78, 119–121, 149–152
Boeing B-29 Superfortress *Bockscar*, 126–127
Boeing B-29 Superfortress *Enola Gay*, 118–127
Bogardus, George, 130–133
Boyd, Albert, 139
Brabazon, Jesse, 21
Breitling Orbiter 3 balloon, 75
Briscoe, Powell, 58
British Aircraft Corporation, 192
British Airways, 194–196
British Overseas Airways Corporation (BOAC), 150
Brown, Arthur, 37
Brown, Russell J., 146
Buck, Kernahan, 68
Buck, Rinker, 68
Bud Light Spirit of Freedom, 75
Byrd, Richard E., 55

Carnegie Institute, 21
Caron, Robert, 126
Caudron G.2, 31–32
Caudron G.3, 32
Caudron G.4, 30–35
Caudron, Gaston, 31
Caudron, René, 31
Central Intelligence Agency (CIA), 182, 201–202
Chicago International Aviation Meet, 18
Chilstrom, Ken, 145
circumnavigation of globe, 37–43, 58–60, 75
Civil Aeronautics Agency (CAA), 130–133
Civil Air Patrol (CAP), 68
Civilian Pilot Training Program (CPTP), 67
Clark, John, 75
Cochran, Jacqueline "Jackie," 145
Cold War, 177, 181
Collin, Frederick, 24
Coolidge, Calvin, 45, 51, 55, 130
Cornish, Joseph J., III, 171–172
Craige, Lawrence G., 106–107
Crossfield, Scott, 165–166

Cunningham, John, 149
Curtiss D-III Headless Pusher, 27–29
Curtiss, Glenn Hammond, 19, 27
Curtiss SB2C Helldiver, 87–88
Curtiss-Wright CW-1 Junior, 65

Dana, Bill, 165
de Havilland Company, 149–150
de Havilland DH 106 Comet 1, 149
de Havilland Ghost 50 turbojet, 149
Defense Advanced Research Projects Agency (DARPA), 200
directional gyro, 59
dive-bombers, 83–89
Domenjoz, John, 26–27
Doolittle, James "Jimmy," 95
Double Eagle II balloon, 75
Douglas Aircraft Company, 38, 79–80, 85
Douglas D-558-1 Skystreak, 145, 164
Douglas DC-1, 79–80
Douglas DC-2, 80
Douglas DC-3, 76–81
Douglas DC-8, 152
Douglas, Donald, 79, 85
Douglas SBD Dauntless, 82–89
Douglas World Cruiser DWC-2 *Boston*, 38, 41–42
Douglas World Cruiser DWC-2 *Chicago*, 36–43
Douglas World Cruiser DWC-2 *New Orleans*, 38, 42–43
Douglas World Cruiser DWC-2 *Seattle*, 38–39
drones, 198–203
Dryden, Hugh L., 74

Earhart, Amelia, 53–57, 61
Echols, Oliver P., 102
Eisenhower, Dwight, 181–182
Elder, Ruth, 55
electronic countermeasures (ECM) systems, 185–186
Enola Gay, 118–127
Eustace, Alan, 75
Excelsior III balloon, 74
Experimental Aircraft Association (EAA), 133
Explorer I balloon, 72–73
Explorer II balloon, 71–75

Ferebee, Thomas, 123, 125

Flynn, S. C., 158–159
Fokker aircraft company, 46
Fokker F.10A, 78
Fokker F.VII *Friendship*, 55
Foulois, Benjamin D., 16
Fowler, Robert, 19, 21
Frye, Jack, 78–79

Galland, Adolf, 111, 116–117
Garber, Paul E., 27, 29, 43, 51
Gatty, Harold, 58–60
Gavin, James, 177
General Atomics MQ-1C Gray Eagle, 203
General Atomics MQ-1L Predator, 198–203
General Atomics MQ-9 Reaper, 200, 203
General Atomics RQ-1 Predator, 201
General Electric Company (GE), 102–103
Gentile, Don, 96
Gloster E.28/39, 101
GNAT 750, 201
Goff, Hal, 69
Gold Cup hydroplane races, 151
Goodlin, Chalmers "Slick," 139
Gordon, Lou, 55
Graf Zeppelin, 58–59
Grahame-White, Claude, 24
Gray, Hawthorne C., 72
Greene, Larry P., 144
Grosvenor, Gilbert, 72
Guest, Amy Phipps, 54–55
Guggenheim, Harry, 64

Hall, Donald A., 46, 48
Hall, F. C., 58
Hall, William E., 83, 86
Harper, Harry, 24
Harvey, Alva, 39
Heath Parasol, 64
Heinemann, Edward, 84
Heinkel, Ernst, 110
Heinkel He 280, 110
helicopters, 175–179
Heron, Samuel, 47
Herrick, Myron, 49
Herring-Curtiss Company, 27
Hinton, Bruce H., 147
Hiroshima, 119, 125, 127
Hitler, Adolf, 115–116
homebuilt airplanes, 129–133
Hoover, Herbert, 60
Hoover, Robert A. "Bob," 139–140
Horkey, Ed., 144
Howard, James H., 96
Howze, Hamilton, 177–178
Hughes, Howard, 80

hypersonic flight, 161–162, 164–166, 183, 192–193.
    See also sound barrier

ILC Dover Industries, 75
International Safe Aircraft Competition, 64–65

Jacobs, Jim, 12
Jamouneau, Walter, 66, 69
Jeffries, John, 72
Jeppson, Morris R., 125
jet engine technology, 101, 104–105, 109–110, 143, 149
John, W. H., 158–159
Johnson, Clarence "Kelly," 107, 182
Johnson, Richard, 170
Johnston, Alvin "Tex," 151

Karem, Abraham, 200–201
Kennedy, John F., 156–157
Kepner, William E., 72–73, 95
Kindelberger, James H. "Dutch," 79, 92
Kittinger, Joseph, 74–75
Kitty Hawk Flyer, 9–13
Knight, William "Pete," 161, 166
Korean War, 68, 97–98, 145–147, 176
Kotcher, Ezra, 136–137

Lahm, Frank P., 16
Land, E. M., 182
Langley, Samuel P., 12–13
Latham, Hubert, 25
LeMay, Curtis E., 122
Lewis, Robert A., 123, 126–127
Lilienthal, Otto, 9–10
Lindbergh, Charles A., 45–49, 51
Lindner, Gerd, 117
Lindsay, David B., 99
Lippisch, Alexander, 110
Little Gee Bee, 128–133
Lockheed A-12, 183–184
Lockheed Aircraft, 53, 182, 189
Lockheed P-38 Lightning, 94
Lockheed P-80 Shooting Star, 107, 146
Lockheed SR-71A Blackbird, 180–189
Lockheed U-2 reconnaissance aircraft, 181
Lockheed Vega 5B, 53–57
Lockheed Vega 5C *Winnie Mae*, 57–61
Long, Les, 131

MacArthur, Douglas, 146
MacRobertson Trophy Air Race, 80
Mantz, Paul, 98
Martin, Frederick L., 38–39
McClure, Clifton, III, 171
McDonnell Aircraft Company, 155–156

McDonnell F-4S-44 Phantom II, 154–159
McDonnell F3H-G Demon, 155
McNamara, Robert, 156–157
McRae, Alexander "Jack," 132
Mercer, Jeff, 133
Meredith effect, 92
Messerschmitt Me 163 Komet, 110
Messerschmitt Me 262 1-a Schwalbe, 108–117, 144
Messerschmitt, Willy, 109
Meyer, John C., 96
Mikoyan-Gurevich MiG-15, 146
Mutke, Hans Guido, 117

Nagasaki, 119, 126–127
National Advisory Committee for Aeronautics
    (NACA), 84, 103, 135, 140–141, 162
National Air Races, 58, 98, 129, 145
National Championship Air Races, 98
National Geographic Society, 72–73
Nebraska Aircraft Corporation, 45
Nelson, Erik H., 38
Nevada Airlines, 58
Niemi, Leonard A., 170, 171
Nieuport scout, 27
Noonan, Fred, 55
North American Aviation (NAA), 92–94, 143, 161
North American F-86A Sabre, 142–147
North American P-51D Mustang, 90–99
North American X-15, 160–167
Northcliffe Prize, 23, 25
Northrop Corporation, 84
Northrop, John, 53
Nott, Julian, 75
Noyes, Blanche W., 55

Ogden, Henry, 41
Orteig Prize, 46

Palmer, James, 179
Pan American Airways, 51
Pan American World Airways, 152
Paragon, 75
Paris Automobile Salon (1908), 24
Parker, Alvin H., 169, 172–173
Parker, Stephen, 173
Parsons, Deak, 125
Patrick, Mason M., 37
Pégoud, Adolphe, 25
Piantanida, Nick, 74
Piccard, Auguste, 72
Pietenpol Air Camper, 64
Piper Aircraft, 99
Piper E-2 Cub, 65–66
Piper J-3 Cub, 62–69
Piper J-4 Coupe, 68

Piper J-5 Cruiser, 68
Piper PA-11 Cub Special, 68
Piper PA-12 Super Cruiser *City of The Angels*, 68
Piper PA-12 Super Cruiser *City of Washington*, 68
Piper Super Cub, 69
Piper, William, 63–67, 69
pitch control, 10
Polaroid Corporation, 182
Post, Wiley, 53, 57–61
Powers, Francis Gary, 181
Poyer, Harland M., 102
Preddy, George E., Jr., 91, 96
pressurized flight suits, 61, 74–75
Project StratEx, 75
Putnam, George P., 55

Randall, John J., 171–172
Raspet, August, 170
Raymond, Arthur, 79
Reaction Motors Inc. (RMI), 137–138, 164
Red Bull Stratos balloon, 75
Red Devil biplane (Baldwin), 27
remote-control weapons systems, 120
remotely piloted aircraft systems (RPASs), 198–203
Republic P-47 Thunderbolt, 94
restoration efforts
    Boeing B-29 Superfortress *Enola Gay*, 127
    Caudron G.4, 35
    Curtiss D-III Headless Pusher, 27–29
    Little Gee Bee, 133
    Messerschmitt Me 262 1-a Schwalbe, 114–115
    Wright Flyer, 13
retractable landing gear, 84
Rich, Ben, 189
Ridley, Jack L., 139–140
Robertson Aircraft Corporation, 46
Rodgers, Calbraith Perry, 17–21
Rodgers, John, 21
Rodgers, Mabel, 21
Rogers, Will, 60–61
Roosevelt, Franklin D., 60
Ross, Harlan, 170
Ross-Johnson RJ-5 sailplane, 170
Rushworth, Robert, 166
Russell, Frank, 16
Rutledge, Tom, 48
Ryan Airlines, 46
Ryan M-2, 46
Ryan Model 147 reconnaissance drone, 202
Ryan NY-P *Spirit of St. Louis*, 45–51

Sage Cheshire Aerospace Inc., 75
sailplanes, 169–173

Saulnier, Raymond, 24
Schmued, Edgar, 92
seatbelts, 25
Selfridge, Thomas E., 16
Shaffer, Frank, 21
Smith, C. R., 80
Smith, Lowell H., 38–39, 43
Soaring Society of America, 131, 171
Société des Phares Blériot, 24
Société Pour L'Aviation et ses Dérivés (SPAD), 27
sodium-cooled exhaust valves, 47
sound barrier, 117, 135–141, 145, 162. *See also* hypersonic flight
Sperry artificial horizon, 59
Spinetta, Jean Cyril, 197
SSTs, 191–197
Stack, John, 136, 141
Stanley, Robert M., 101, 105–107
Stevens, Albert W., 71–74
Stiborik, Joe, 126
Stidd, Robert, 179
Story, Tom, 131–132
Strato-Jump III balloon, 74
streamlining, 53
Stultz, Wilmer, 55
Sud Aviation, 192
Sweeney, Charles W., 126
swept wings, 144

Tata Institute of Fundamental Research, 75
Taylor Brothers Aircraft Company, 64
Taylor, Charles, 11, 19, 21
Taylor, Clarence Gilbert, 63–66
Taylor Company, 64–65
Taylor, Gordon, 64
Taylor J-2 Cub, 66
Taylorcraft Model A, 66
Thaden, Louise M., 55
Thompson, Art, 75
1,000-kilometer flights, 169
Tibbetts, Paul W., 123–127
Tibbs, Burrell, 58
Trans Florida Aviation, 99
Trans World Airlines (TWA), 51, 78–79
Transcontinental Air Transport, 51
Transcontinental and Western Air, 78
Trippe, Juan, 152
Tupolev Tu-104, 150
Turcat, André, 192
Turner, Roscoe, 58

United Air Lines, 78
United Parachute Technologies, 75
United States National Soaring Championships, 169

unmanned aerial vehicles (UAVs), 198–203

Van Kirk, Theodore "Dutch," 123
Vandenberg, Hoyt S., 141
Védères, Alice, 24
Vejtasa, Stanley "Swede," 86
Vickers Vimys, 37
Vietnam War, 157–159, 175–176, 178
Voisin Type 8, 35
Vultee, Gerard, 53

WAC Corporal rocket, 162
Wade, Leigh, 38, 41
Walcott, Charles, 35
Walker, Joe, 161, 166
Welch, George, 145
Wetmore, Alexander, 107, 141
Whittle, Frank, 102, 104
Wiggin, Charles, 21
Wiggin, Mabel, 21
wind tunnels, 10–11, 103, 162
wing area, 10
wing curvature, 10
wing warping, 24–25
Women Airforce Service Pilots (WASPs), 68, 107
Woods, Robert J., 137
Woolams, Jack, 138–139
World Trade Center attacks, 200
World War I, 24, 31–34
World War II, 13, 63, 67–68, 74, 77, 81, 83–89, 91–97, 109–117, 119–127, 176
Wright Company, 16, 19
Wright EX *Vin Fiz*, 17–21
Wright Flyer, 9–13
Wright Military Flyer, 14–17
Wright, Orville, 9–13, 16, 19, 27
Wright, Wilbur, 9–13, 16, 19, 24–25, 27
Wyrock, Bob, 133

Yeager, Charles E. "Chuck," 139–141, 145, 162
Yom Kippur War, 157

Quarto is the authority on a wide range of topics.

Quarto educates, entertains and enriches the lives of
our readers—enthusiasts and lovers of hands-on living.

www.quartoknows.com

Compilation © 2016 Quarto Publishing Group USA Inc.
Text © 2016 Smithsonian Institution

First published in 2016 by Zenith Press, an imprint of Quarto Publishing Group USA Inc., 400 First Avenue North, Suite 400,
Minneapolis, MN 55401 USA. Telephone: (612) 344-8100 Fax: (612) 344-8692

quartoknows.com
Visit our blogs at quartoknows.com

Zenith Press titles are also available at discounts in bulk quantity for industrial or sales-promotional use. For details contact the
Special Sales Manager at Quarto Publishing Group USA Inc., 400 First Avenue North, Suite 400, Minneapolis, MN 55401 USA.

10 9 8 7 6 5 4 3 2 1

ISBN: 978-0-7603-5027-0

Library of Congress Cataloging-in-Publication Data

Names: Van der Linden, F. Robert, author. | Spencer, Alex M., author. |
 Paone, Thomas J., author.
Title: Milestones of flight : the epic of aviation with the National Air and
 Space Museum / F. Robert van der Linden, Alex M. Spencer, Thomas J. Paone.
Description: Washington, D.C. : National Air and Space Museum, in association
 with Zenith Press, 2016. | Includes index.
Identifiers: LCCN 2016000000 | ISBN 9780760350270 (plc w/ jacket)
Subjects: LCSH: Aeronautics--History. | Airplanes--Pictorial works. |
 National Air and Space Museum--Catalogs.
Classification: LCC TL515 .V25 2016 | DDC 629.130973--dc23
LC record available at http://lccn.loc.gov/2016000000

Acquiring Editor: Erik Gilg
Project Managers: Madeleine Vasaly and Dennis Pernu
Art Direction: James Kegley
Layout: Kim Winscher

Front cover: The *Spirit of St. Louis* (NASM 2001-2949). *NASM photo by Eric Long*
Case: A detail view of the control jets on the X-15 fuselage (NASM 2007-12580). *NASM photo by Eric Long*
Endpapers: A cockpit view of the Douglas DC-3 shows the instrument panel
          and pilot and co-pilot controls. *Courtesy of Boeing Images*

Printed in China